Restless Multi-Armed Bandit in Opportunistic Scheduling

Kehao Wang • Lin Chen

Restless Multi-Armed Bandit in Opportunistic Scheduling

Springer

Kehao Wang
Wuhan University of Technology
Wuhan, China

Lin Chen
Sun Yat-sen University
Guangzhou, Guangdong, China

ISBN 978-3-030-69961-1 ISBN 978-3-030-69959-8 (eBook)
https://doi.org/10.1007/978-3-030-69959-8

This Springer imprint is published by the registered company Springer Nature Switzerland AG
The registered company address is: Gewerbestrasse 11, 6330 Cham, Switzerland

Preface

A colleague of high repute asked an equally well-known colleague:

– *What would you say if you were told that the multi-armed bandit problem had been solved?*
– *Sir, the multi-armed bandit problem is not of such a nature that it can be solved."*

Story told by Peter Whittle

Restless Multiarmed Bandit (RMAB) has been a classical problem in stochastic optimization and reinforcement learning with a wide range of engineering applications, including but not limited to, multi-agent systems, web search, Internet advertising, social networks, queueing systems, etc. In this book, we present a systematic research on a number of fundamental problems related to performance and computing complexity of both myopic policy and index policy under the context of imperfect sensing or observation appeared in practical scenarios. These problems are well characterized in mathematics and intuitively understandable, while of both fundamental and practical importance, and require nontrivial effort to solve. Especially, we address the following problems ranging from theoretical analysis to practical policy implementation and optimization:

- Sufficient conditions under which the myopic policy is optimal for homogeneous two-state Markov channels
- Feasibility and computation of the index policy for heterogeneous two-state Markov channels
- Sufficient conditions under which the myopic policy is optimal for homogeneous multistate Markov channels
- Feasibility and computation of the index policy for heterogeneous multistate Markov channels

Actually, seeking the optimal policy of a generic RMAB involves not only the tradeoff between exploration and exploitation, but also the balance between aggression and conservation which are important concepts in machine learning. In this book, we adopt a search and exposition line from theoretical analysis to practical policy implementation and optimization.

To lay the theoretical foundations for the design and optimization of the policies with linear complexity, we start by investigating the basic concepts and obstacles of RMAB. Technically, a unified framework is constructed under the myopic policy in terms of the regular reward function, characterized by three basic axioms—symmetry, monotonicity, and decomposability. For the homogeneous channels, we establish the optimality of the myopic policy when the reward function can be expressed as a regular function and when the discount factor is bounded by a closed-form threshold determined by the reward function.

In order to further obtain the asymptotically optimal performance of the RMAB in a larger parameter space, we analyze the feasibility and computation scheme of the Whittle index for two-state Markov case by the fixed-point approach. We first derive a threshold structure of the single-arm policy. Based on this structure, the closed-form Whittle index is obtained for the case of negatively correlated channels, while the Whittle index for the positively correlated channel is hugely complicated for its uncertainty, particularly for certain regions below stationary distribution of Markov chain. Then, this region is divided into the deterministic regions and the indeterministic regions with interleaving structure. In the deterministic regions, the evolution of the dynamic system is periodic and then there exists an Eigen matrix to depict this kind of evolving structure through which the closed-form Whittle can be derived. In the indeterministic regions, there does not exist an Eigen matrix to depict its aperiodic structure. In the practical scenarios, given the computing precision, the Whittle index in those regions can be computed in the simply linear interpolation since the distribution of the deterministic and indeterministic regions appears in the interleaving form.

We further consider an opportunistic scheduling system consisting of multiple homogeneous channels evolving as a multistate Markov process. We carry out a theoretic analysis on the performance of myopic policy with imperfect observation, introduce monotonic likelihood ratio in order to characterize the evolving structure of belief information, and establish a set of closed-form conditions to guarantee the optimality of the myopic scheduling policy in the opportunistic scheduling system.

For the heterogeneous case, we cast the problem to a restless bandit. The pivot to solve restless bandit by index policy is to establish feasibility of the index policy or indexability. Despite the theoretical and practical importance of the index policy, the indexability is still open for the opportunistic scheduling in the heterogeneous multistate channel case. To fill this gap, we mathematically propose a set of sufficient conditions on channel state transition matrix under which the indexability is guaranteed, and consequently, the index policy is feasible. We further develop a simplified procedure to compute the index by reducing the complexity from more than quadratic to linear. Our work consists of a small step toward solving the opportunistic scheduling problem in its generic form involving multistate Markov channels.

Wuhan, China Kehao Wang
Guangzhou, China Lin Chen
January 9, 2021

Acronyms

eFOSD	Extended first-order stochastic dominance
FOSD	First-order stochastic dominance
i.i.d.	Independent and identically distributed
MAB	Multiarmed bandit
MLR	Monotone likelihood ratio
MPI	Marginal productivity index
PB	Proportionally best
POMDP	Partially observable markov decision process
RB	Relatively best
RMAB	Restless multiarmed bandit
RMABs	Restless multiarmed bandits
SB	Scored based
TP2	Totally positive of order 2

Contents

Chapter 1
RMAB in Opportunistic Scheduling

1.1 Introduction

1.1.1 Multiarmed Bandit

Multiarmed bandit, first posed in 1933 for clinical trial, has become a classical problem in stochastic optimization and reinforcement learning with a wide range of engineering applications, including but not limited to, multi-agent systems, web search, Internet advertising, social networks, and queueing systems.

Consider a dynamic system consisting of a player and N independent arms. In each time slot t ($t = 1, 2, \ldots$), the state of arm k is denoted by $s_k(t)$ and completely observable to the player. At slot t, the player selects one arm, i.e., arm k, to activate based on the system state $\mathcal{S}(t) = [s_1(t), s_2(t), \cdots, s_N(t)]$ and accrues reward $R(s_k(t))$ determined by the state $s_k(t)$ of arm k. Meanwhile, the state of arm k will transit to another state in the next slot according to certain transition probabilities, i.e., $p_{i,j}^{(k)} = P(s_k(t+1) = j \mid s_k(t) = i), i, j \in \Omega_k$, where Q_k denotes the state space of arm k. The states of other arms which are not activated will remain frozen, i.e., $s_n(t + 1) = s_n(t)\ \forall\ n \neq k$.

The player's selection policy $\pi = \{\pi(1), \pi(2), \cdots\}$ is a serials of mapping from the system state $\mathcal{S}(t)$ to the action $a(t)$ indicating which arm is activated, i.e., $\pi(t) : \mathcal{S}(t) \mapsto a(t)$. The objective is to obtain the optimal policy π^* to maximize the expected total discounted reward in an infinite horizon:

K. Wang, L. Chen, *Restless Multi-Armed Bandit in Opportunistic Scheduling*,
https://doi.org/10.1007/978-3-030-69959-8_1

$$\pi^* = \arg\max_{\pi} \mathbb{E}\left[\lim_{T\to\infty} \sum_{t=1}^{T} \beta^{t-1} R(s_{a(t)}(t))\right] \tag{1.1}$$

where the discount factor $0 \leq \beta < 1$.

Since the size of the system states grows exponentially with the number of arms, the above problem, called the classic MAB problem, has an exponential complexity for its general numerical solutions.

This sequential decision problem (1.1) was firstly proposed by Thompson in 1933 [1], but the theoretical structure of the optimal solution for the classic MAB has not been obtained until Gittins's seminal work [2] in 1974. Gittins showed that an index policy is optimal, called Gittins index later, and thus reduces the complexity of the problem from exponential to linear with the number N of arms.

Theorem 1.1 (Gittins, 1974) *The optimal policy has an index form. Specially, for all $1 \leqslant k \leqslant N$, there exists an index function $Gk(\bullet)$ that maps the state $i \in \Omega_k$ of arm k to a real number. At each time, the optimal action is to activate the arm with the largest index.*

Gittins also gave a specific form of the index function $Gk(\bullet)$, referred as Gittins index, as given in the following definition.

Definition 1.1 (Gittins Index) For any state $i \in \Omega_k$ of arm k,

$$G_k(i) = \limsup_{\sigma \geqslant 1} \frac{\mathbb{E}\left[\sum_{t=1}^{\sigma} \beta^{t-1} \mathcal{R}(s_k(t)) \mid s_k(1) = i\right]}{\mathbb{E}\left[\sum_{t=1}^{\sigma} \beta^{t-1} \mid s_k(1) = i\right]} \tag{1.2}$$

where σ is a stopping time for activating the arm k.

Basically, Gittins index measures the maximum reward rate that can be achieved by focusing on activating one arm starting from its current state. Therefore, by Gittins index, the player can accrue reward as quickly as possible and thus maximize the total discounted reward.

1.1.2 Restless Multiarmed Bandit

P. Whittle [3] extended the MAB to a more general model in which a set of K ($K > 1$) arms, denoted as $\mathcal{K}(t)$, can be activated simultaneously and change their states in each slot and meanwhile the passive arms are also allowed to offer reward and change state, which makes it different from the classic MAB.

If arm k is activated, then its state transits according to a transmitting rule P_{k1} and yields the immediate reward $g_{k1}(s_k(t))$ while it transits by another rule P_{k2} and yields the immediate reward $g_{k2}(s_k(t))$ when arm k is not activated. A policy $\pi = \{\pi(t)\}_{t=1}^{\infty}$ is a serial of mappings where π_t maps the system state $S(t)$ to the set of K arms $\mathcal{K}(t)$ to be activated in slot t.

In [3], P. Whittle considered the above problem to maximize the average reward over an infinite horizon,[1] which can be formulated as follows:

$$\pi^* = \arg\max_{\pi} \mathbb{E}\left\{ \lim_{T\to\infty} \frac{1}{T} \sum_{t=1}^{T} \left[\sum_{i\in\mathcal{K}(t)} g_{i1}(s_i(t)) + \sum_{j=1, j\notin\mathcal{K}(t)}^{N} g_{j2}(s_j(t)) \right] \right. . \tag{1.3}$$

Let γ_k denote the maximum expected average reward obtained by playing arm k without constraint:

$$\gamma_k = \max_{\pi} \mathbb{E}\left[\lim_{T\to\infty} \frac{1}{T} \sum_{t=1}^{I} g_{ka_k(t)}(s_k(t)) \right], \text{where } a_k(t) \in \{1,2\} \tag{1.4}$$

Let $f_k(s_k(1))$ denote the differential reward caused by the transient effect of starting from state $s_k(1)$ rather than from an equilibrium situation:

$$f_k(s_k(1)) = \lim_{T\to\infty} \mathbb{E}_{\pi^*} \left[\frac{1}{T} \sum_{t=1}^{T} g_{ka_k(t)}(s_k(t)) - \gamma_k \right] \tag{1.5}$$

We have the following optimal equation for the maximum expected average reward γ_k:

$$\gamma_k + f_k(s_k(t)) = \max_{a=\{1,2\}} [g_{ka}(s_k(t)) + \mathbb{E}[f_k(s_k(t+1)) \mid s_k(t)]] \tag{1.6}$$

We can rewrite the above formulation more compactly as

$$\gamma_k + f_k(s_k(t)) = \max\,[L_{k1}f_k, L_{k2}f_k] \tag{1.7}$$

We consider the following relaxed condition: K out of N arms are activated on average rather than exactly in all time slots, i.e.,

$$\mathbb{E}[|\mathcal{K}(t)|] = K \text{ instead of } \mid \mathcal{K}(t) \mid = K, \forall t \tag{1.8}$$

Then the objective under the relaxed condition is the following:

[1] The discounted reward can be similarly discussed.

$$\max \mathbb{E}\left[\sum_{n=1}^{N} r_n\right], \text{s.t.} \mathbb{E}\left[\sum_{n=1}^{N} I_n\right] = K \tag{1.9}$$

where r_n is the average reward obtained from arm n under the relaxed constraint, and $I_n = 1; 0$ according to whether arm n is activated or not.

We have the objective by the classic Lagrangian multiplier as follows:

$$\max \mathbb{E}\left[\sum_{n=1}^{N} r_n + v\sum_{n=1}^{N} I_n\right] = \max \mathbb{E}\left[\sum_{n=1}^{N} (r_n + vI_n)\right] \tag{1.10}$$

We thus have a v-subsidy problem

$$\gamma_k(v) + f_k = \max\left[L_{k1} f_k, v + L_{k2} f_k\right] \tag{1.11}$$

where v is referred as subsidy for passivity.

We define the index $W_k(i)$ of arm k in state $i \in \Omega_k$ as the value of v which makes the active and the passive phases equally attractive:

$$L_{k1} f_k = v + L_{k2} f_k \tag{1.12}$$

Let $\mathcal{P}_k(v)$ be the set of states for which arm k would be passive under a v-subsidy policy. Then the arm is indexable if $\mathcal{P}_k(v)$ increases monotonically from $\varnothing$ to Ω_k as V increases from $-\infty$ to $+\infty$ ..

Thus, if all arms are indexable, arm k will be activated in slot t if $W_k(s_k(t)) > v$. Therefore, we obtain the following Whittle index policy.

Definition 1.2 (Whittle Index Policy) If all the bandits are indexable, activate the K arms of the greatest indices in each slot.

Conjecture 1.1 (Whittle Conjecture) Suppose all arms are indexable, the index policy is optimal in terms of average yield per arm in the limit.

1.2 Technical Challenge

For MAB or RMAB, one research thrust of existing literatures focuses on seeking sufficient conditions under which myopic or greedy policy which only maximizes the current slot reward is optimal [4–7]. The second one is to study the asymptotically optimal Whittle index policy [8–18]. The third one is to derive some application-oriented approximate heuristic policies. However, these works do not consider the following three main challenges.

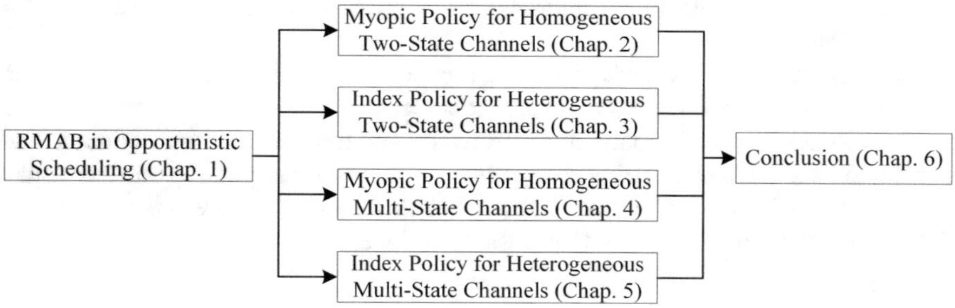

Fig. 1.1 Book organization

- Partial Information: In an opportunistic scheduling system, the decision-maker or scheduler has to consume a certain resource (i.e., time, energy, frequency) to observe (or sense, detect, sample) the system state. As a result, the decision-maker would not or cannot observe the complete state of the system due to the practical cost constraint of consuming resources, and can only obtain partial information of the system state. Thus, based on the partial information, the scheduler has to make decision by learning decision history and observation history.
- Imperfect Information: In practical environment, the decision-maker needs relying on a certain appliance to obtain the system state, and consequently, cannot observe the perfect information since any appliance would bring a certain mistake, i.e., false alarm and missing alarm. Hence, the imperfect observation is unavoidable during the process of obtaining information in a scheduling system. This kind of imperfect information leads to complicated nonlinear dynamics of the system, which requires special technique to conquer.
- Multi-State Information: Actually, to make a better decision for an opportunistic scheduling system, the scheduler needs to know more precise system information. Thus, it is required to characterize the system state in a small grain level rather than a simply macro one, such as two-state (good vs. bad, or 1 vs. 0) using only one threshold. Multi-threshold or multi-state is adopted to describe the system state in a small grain level. However, multi-state quantity provably requires multi-variance technique.

1.3 Book Organization

In this book, we adopt a research and exposition line from theoretical modeling and analysis to practical algorithm design and optimization. Figure 1.1 illustrates the structure of the book. In the remainder of this section, we provide a high-level overview of the technical contributions of this book, which are presented sequentially in Chaps. 2–5.

To facilitate readers, we adopt a modularized structure to present the results such that the chapters are arranged as independent modules, each devoted to a specific topic outlined above. In particular, each chapter has its own introduction and conclusion sections, elaborating the related work and the importance of the results with the specific context of that chapter. For this reason, we are not providing a detailed background, or a survey of prior work here.

References

1. W.R. Thompson, On the likelihood that one unknown probability exceeds another in view of the evidence of two samples. Biometrika **25**(3/4), 285–294 (1933)
2. J.C. Gittins, D.M. Jones, A dynamic allocation index for the sequential design of experiments. Prog. Statist, 241–266 (1974)
3. P. Whittle, Restless bandits: Activity allocation in a changing world. J. Appl. Prob. **25A**, 287298 (1988)
4. F.E. Lapiccirella, K.Q. Liu, Z. Ding, Multi-channel opportunistic access based on primary ARQ messages overhearing. Proc. IEEE ICC, 1–5 (2011)
5. Q. Zhao, B. Krishnamachari, K.Q. Liu, On myopic sensing for multi-channel opportunistic access: Structure, optimality, and performance. IEEE Trans. Wirel. Commun. **7**(3), 5413–5440 (2008)
6. S. Ahmand, M.Y. Liu, T. Javidi, Q. Zhao, B. Krishnamachari, Optimality of myopic sensing in multichannel opportunistic access. IEEE Trans. Inf. Theory **55**(9), 4040–4050 (2009)
7. S. Murugesan, P. Schniter, N.B. Shroff, Multi-user scheduling in makov-modeled downlink using randomly delayed ARQ feedback. IEEE Trans. Inf. Theory **58**(2), 1025–1042 (2012)
8. H. Ji, C.V. Leung, C.Q. Luo, F.R. Yu, Optimal channel access for tcp performance improvement in cognitive radio networks. Wirel. Netw **17**, 479–492 (2010)
9. D. Chen, H. Ji, X. Li, Distributed best-relay node selection in underlay cognitive radio networks a restless bandits approach. Proc. IEEE WCNC, 1208–1212 (2011)
10. M.Y. Liu, N. Ehsan, On the optimality of an index policy for bandwidth allocation with delayed state observation and differentiated services. Proc. IEEE INFOCOM **3**, 1974–1983 (2004)
11. P. Jacko, Value of information in optimal flow-level scheduling of users with markovian timevarying channels. Perform. Eval. **68**(11), 1022–1036 (2011)
12. K.Q. Liu, Q. Zhao, Indexability of restless bandit problems and optimality of whittle index for dynamic multichannel access. IEEE Trans. Inf. Theory **56**(11), 5547–5567 (2000)
13. J.L. Ny, M. Dahleh, E. Feron, Multi-uav dynamic routing with partial observations using restless bandit allocation indices. Proc. ACC, 4220–4225 (2008)
14. O. Jonathan, A continuous-time markov decision process for infrastructure surveillance. Oper. Res. Proc., 327–332 (2010)

15. B. Sanso, P. Jacko, Optimal anticipative congestion control of flows with time-varying input stream. Perform. Eval. **69**(2), 86–101 (2012)
16. V. Raghunathan, V. Borkar, M. Cao, P.R. Kumar, Index Policies for Real-Time Multicast Scheduling for Wireless Broadcast Systems. Proc. IEEE INFOCOM, 2243–2251 (2008)
17. T. He, A. Anandkumar, D. Agrawal, Index-based sampling policies for tracking dynamic networks under sampling constraints. Proc. IEEE INFOCOM, 1233–1241 (2011)
18. U. Ayesta, E. Martin, P. Jacko, A modeling framework for optimizing the flow-level scheduling with time-varying channels. Perform. Eval. **67**(11), 1024–1029 (2010)

Chapter 2
Myopic Policy for Opportunistic Scheduling: Homogeneous Two-State Channels

2.1 Introduction

2.1.1 Background

We consider an opportunistic multichannel communication system where a user can access multiple Gilbert-Elliot channels, but it is limited to sense and transmit on a subset of channels each time. The fundamental problem that we are interested in is how the user can exploit past observation history, past decision history, and the knowledge of the stochastic properties of those channels to maximize his/her utility (e.g., expected throughput) by switching channels in each decision period opportunistically.

Formally, there are N i.i.d. channels, each evolving as a two-state Markov process where the state of a channel indicates the desirability of accessing this channel. At each time slot, the user chooses k $(1 < k < N)$ of the N channels to sense, access, and obtain a certain amount of reward which depends on the states of the chosen channels. Given the initial state of the system, i.e., the initial states of N channels, the goal of the user is to find the optimal policy to schedule channels at each time slot so as to maximize the accumulated discounted reward. This channel access problem can be cast into the RMAB problem [1] or partially observable Markov decision process (POMDP) [2].

2.1.2 Related Work

Due to its application in numerous engineering problems, the RMAB problem is of fundamental importance in stochastic decision theory. However, finding the optimal policy for the generic RMAB problem is shown to be PSPACE-hard by Pa-padimitriou et al. [3]. P. Whittle in [1] proposed a heuristic index policy called

K. Wang, L. Chen, *Restless Multi-Armed Bandit in Opportunistic Scheduling*,
https://doi.org/10.1007/978-3-030-69959-8_2

Whittle index policy which is shown to be asymptotically optimal in certain limited regime under some specific constraints [4]. In this regard, Liu et al. studied in [5] the indexability of a class of RMAB problems relevant to dynamic multichannel access applications. However, the optimality of the index policy based on Whittle approach is not guaranteed in general cases, especially when the channels follow heterogeneous Markov chains. Moreover, not every RMAB problem has a well-defined Whittle index.

A natural alternative, given the intractability of the RMAB problem, is to seek a simply myopic policy maximizing the short-term reward, i.e., slot reward. In this line of research, significant research efforts have been devoted to studying the performance of the myopic policy, especially in the context of opportunistic spectrum. Some key contributions from recent works on this subject can be summarized as follows. Zhao et al. [6] established the structure of the myopic sensing policy, analyzed the performance, and partly obtained the optimality for the case of homogeneous i.i.d. channels. Ahmad and Liu et al. [7] derived the optimality of the myopic policy for the positively correlated homogeneous i.i.d. channels when the user is limited to access one channel at each time slot. Ahmad and Liu [8] further extended the optimality result to the case of sensing multiple homogenous channels ($k > 1$) for a particular form of utility function which is used to model the fact that the user gets one unit of reward for each channel sensed to be good. Our works studied the case of non i.i.d. channels and provided generic conditions on reward function under which the myopic policy is optimal [9], and also illustrated that when these conditions are not satisfied, the myopic policy may not be optimal [10].

2.1.3 Main Results and Contributions

The vast majority of previous works (i.e., [2, 6–9]) in the area assume that the user can achieve perfect observation of channel state. However, sensing or observation errors are inevitable in practical scenario due to noise and system hardware limitation, especially in the dynamic environment of wireless communication. More specifically, a good (bad, respectively) channel may be sensed as bad (good, respectively). In such an imperfect context, it is crucial to study the structure and the optimality of the myopic sensing policy with imperfect observation. We would like to emphasize that the presence of sensing or observation error brings two obstacles when studying the myopic sensing policy in this new context.

- The belief value of channel evolves as a nonlinear mapping in the imperfect case, instead of a linear one in the perfect case.
- The update of belief value of channel depends not only on the channel Markov evolution rule but also on the observation outcome, impliciting that the transition is not deterministic.

Due to the above particularities, the problem considered in this chapter requires an original study on the optimality of the myopic sensing policy, which cannot draw on existing results in the perfect sensing case. We would like to report that despite its practical importance and particularities, very few works have been done on the impact of sensing error on the performance of the myopic sensing policy, or more generically, on the RMAB problem under imperfect observation.

To the best of our knowledge, references [11, 12] are the only work in this area. Chen and Zhao et al. [11] decoupled the design of the sensing strategy from that of the spectrum sensor and the access strategy, and reduced the constrained POMDP to an unconstrained one. Liu and Zhao and Krishnamachari [12] established the optimality of the myopic policy for the case of two channels with a particular utility function under certain conditions and conjectured the optimality for arbitrary N under the same conditions. In this chapter, we derive closed-form conditions to guarantee the optimality of the myopic sensing policy for arbitrary N and for a class of utility functions. As shown in Sect. 2.4.3, the result obtained in the chapter can cover the result of [12]. Moreover, this chapter also significantly extends our previous work [9], focusing on perfect sensing scenario in which the analysis cannot be applied in the imperfect sensing scenario due to the non-trivial particularities introduced by sensing error as mentioned previously. In this regard, our work in this chapter contributes the existing literature by developing an adapted analysis on the RMAB problem under imperfect sensing.

2.2 Problem Formulation

2.2.1 *System Model*

Table 2.1 summaries main notations used in this chapter.

We consider a multichannel opportunistic communication system, in which a user is able to access a set $\mathcal{N}$ of N homogeneous channels, each characterized by a Markov chain of two states, *good* (1) and *bad* (0). The state transition matrix $\boldsymbol{P}$ of those channels is given as follows:

$$\boldsymbol{P} = \begin{bmatrix} 1-p_{01} & p_{01} \\ 1-p_{11} & p_{11} \end{bmatrix} \tag{2.1}$$

Assume that the system operates in a synchronous time slot fashion with the time slot indexed by $t(t = 0, 1, \ldots, T)$, where T is the time horizon of interest. We assume that these channels go through state transition at the beginning of each slot t.

Due to hardware constraints and energy cost, the user (precisely, the spectrum detector) is practically allowed to sense only $k(1 \le k \le N)$ of the N channels at each time slot t. We assume that the user makes decision on channel selection at the beginning of each time slot after the state transition of these channels. Once a channel is chosen, the user detects the channel state $S_i(t)$, which can be considered as a binary hypothesis test as follows.

Table 2.1 Main notations

Symbols	Descriptions
$\mathcal{N}$	The set of N channels, i.e., 1, 2, . . ., N
$\mathcal{N}(m)$	m channels in N, i.e., 1, 2, . . .,m
$\boldsymbol{P}$	Channel state transition matrix
T	The total number of time slots
t	Time slot index
ε	False alarm probability
ζ	Miss detection probability
$\mathcal{A}$	The set of channel chosen in slot t
$\omega_i(t)$	The conditional probability of being "good"
$\Omega(t)$	Channel state belief vector at slot t
$\Omega(0)$	The initial channel state belief vector
$O_i(t)$	The observation state of channel i
π_t	The mapping from $\Omega(t)$ to $\mathcal{A}(t)$
β	Discount factor
$R(\pi_t(\Omega(t)))$	The reward collected in slot t
$V_t(\Omega(t))$	Value function in slot t

$$\mathcal{H}_0 : S_i(t) = 1(\text{ good }) \text{ vs.} \mathcal{H}_1 : S_i(t) = 0 \text{ (bad)}. \tag{2.2}$$

We assume that the performance of channel state detection is characterized by the probability of false alarm ε and the probability of miss detection ζ:

$$\mathcal{E} \triangleq \Pr\{ \text{ decide } \mathcal{H}_1 \mid \mathcal{H}_0 \text{ is true } \}$$
$$\zeta \triangleq \Pr\{ \text{ decide } \mathcal{H}_0 \mid \mathcal{H}_1 \text{ is true } \}.$$

We denote the set of channels chosen by the user at time slot t by $\mathcal{A}(t)$ where $\mathcal{A}(t) \in \mathcal{N}$ and $\mid \mathcal{A}(t) \mid = k$. Based on the imperfect sensing observations $\{O_i(t) \in \{0, 1\} : i \in \mathcal{A}(t)\}$ in slot t, the user decides whether to access channel i for transmission.

2.2.2 *Restless Multiarmed Bandit Formulation*

Obviously, by imperfectly sensing only k out of N channels, the user cannot observe the state information of the whole system. Hence, the user has to infer the channel states from its past decision and observation history so as to make its future decision.

To this end, we define the *channel state belief vector* (hereinafter referred to as *belief vector* for briefness) $\Omega(t) \triangleq \{\omega_i(t), i \in \mathcal{N}\}$, where $0 \leq \omega_i(t) \leq 1$ is the conditional probability that channel i is in good state (i.e., $S_i(t) = 1$).

Given the sensing action $\mathcal{A}(t)$ and the observations $\{O_i(t) \in \{0, 1\} : i \in \mathcal{A}(t)\}$, the belief vector in $t + 1$ slot can be updated recursively using Bayes Rule as shown in (2.3).

$$\omega_i(t+1) = \begin{cases} p_{11}, & \text{if } i \in \mathcal{A}(t), O_i(t) = 1 \\ \Gamma(\varphi(\omega_i(t))), & \text{if } i \in \mathcal{A}(t), O_i(t) = 0 \\ \Gamma(\omega_i(t)), & \text{if } i \notin \mathcal{A}(t) \end{cases} \tag{2.3}$$

where,

$$\Gamma(\omega_i(t)) \triangleq \omega_i(t) p_{11} + (1 - \omega_i(t)) p_{01} \tag{2.4}$$

$$\varphi(\omega_i(t)) \triangleq \frac{\varepsilon \omega_i(t)}{1 - (1 - \mathcal{E}) \omega_i(t)} \tag{2.5}$$

Remark 2.1 We would like to emphasize that the sensing error introduces further complications in the system dynamics (i.e., $\varphi(\omega)$ is nonlinear in m) compared with the perfect sensing case. Therefore, those results [7, 9] obtained without sensing error cannot be trivially extended to the scenario with sensing error.

A sensing policy specifies a sequence of functions $\pi := [\pi_0, \pi_1, \ldots, \pi_T]$ where $\pi_t (0 \leqslant t \leqslant T)$r maps the belief vector $\Omega(t)$ to the action $A(t)$ at each time slot t:

$$\pi_t : \Omega(t) \mapsto \mathcal{A}(t), \ | \mathcal{A}(t) | = k \tag{2.6}$$

Given the imperfect sensing context, we are interested in the user's optimization problem to find the optimal sensing policy π^* that maximizes the expected total discounted reward over a finite horizon:

$$\pi^* = \arg\max_{\pi} \mathbb{E} \left\{ \sum_{t=0}^{T} \beta^{t-1} R_{\pi_t}(\Omega(t)) \mid \Omega(0) \right\} \tag{2.7}$$

where $R_{\pi_r}(\Omega(t))$ is the reward collected in slot t under the sensing policy π_t with the initial belief vector $\Omega(0)^{1}$,[1] and $0 \leq \beta \leq 1$ is the discount factor characterizing the feature that the future rewards are less valuable than the immediate reward. By treating the belief value of each channel as the state of each arm of a bandit, the user's optimization problem can be cast into a restless multiarmed bandit problem.

2.2.3 *Myopic Sensing Policy*

In order to get more insight on the structure of the optimization problem formulated in (2.7) and the complexity to solve it, we derive the dynamic programming formulation of (2.7) as follows.

---

[1]If no information on the initial system state is available, each entry of $Q(0)$ can be set to the stationary distribution $\omega_0 = \frac{P_{01}}{1 + p_{01} - P_{11}}$.

$$\begin{cases} V_T(\Omega(T)) = \max_{\mathcal{A}(T)} \mathbb{E}\Big[R_{\pi_T}(\Omega(T))\Big] \\ V_t(\Omega(t)) = \max_{\mathcal{A}(t)} \mathbb{E}\Big[R_{\pi_t}(\Omega(t)) + \beta \sum_{\mathcal{E} \subseteq \mathcal{A}(t)} \Pr(\mathcal{A}(t), \mathcal{E}) V_{t+1}(\Omega_{\mathcal{E}}(t+1))\Big] \end{cases} \tag{2.8}$$

where

$$\Pr(\mathcal{A}(t), \mathcal{E}) \triangleq \prod_{i \in \mathcal{E}} (1 - \mathcal{E})\omega_i(t) \prod_{j \in \mathcal{A}(t) \backslash \mathcal{E}} [1 - (1 - \mathcal{E})\omega_j(t)]. \tag{2.9}$$

In the above Bellman Eq. (2.8), $V_t(\Omega(t))$ is the value function corresponding to the maximal expected reward from time slot t to $T(0 \leq t \leq T)$ with the believe vector $\Omega(t + 1)$ following the evolution (2.3), given that the channels in the subset $\mathcal{E}$ are sensed to be good and the channels in $\mathcal{A}(t) \backslash \mathcal{E}$ are sensed to be bad.

Solving (2.7) through the above recursive iteration is computationally heavy due to the fact that the belief vector $\{\Omega(t), t = 0, 1, \cdots, T\}$ is a Markov chain with uncountable state space when $T \to \infty$, resulting in the difficulty in tracing the optimal sensing policy π^*. Hence, a natural alternative is to seek simple myopic sensing policy which is easy to compute and implement that maximizes the immediate reward, formally defined as follows:

Definition 2.1 (Myopic Policy) Let $U(\mathcal{A}(t), \Omega(t)) := \mathbb{E}\{R_{\pi_t}(\Omega(t))\}$ denote the expected immediate reward obtained in time slot t under the sensing policy π_t : $\Omega(t) \mapsto \mathcal{A}(t)$. The myopic sensing policy $\widehat{\mathcal{A}}(t)$, consists of sensing the k channels that maximizes $U(\mathcal{A}(t), \Omega(t))$, i.e.,

$$\widehat{\mathcal{A}}(t) = \max_{\pi_t} \mathbb{E}\{R_{\pi_t}(\Omega(t))\} \tag{2.10}$$

Despite the simple and robust structure of the myopic policy, the optimality of this kind of greedy policy is not guaranteed. More specifically, when the channels are homogeneous (i.e., all channels follow the same Markovian dynamics according to P) and positively correlated (i.e., $p_{11} \geqslant p_{01}$), the myopic sensing policy is shown to be optimal when the user is limited to sensing one channel each slot ($k = 1$) and obtains one unit of reward when the sensed channel is good [6]. The analysis [7] and our previous work [10] further extend the study on the generic case with $k \geq 1$. However, the authors [7] showed that the myopic sensing policy is optimal if the user gets one unit of reward for each channel sensed to be good,[2] while our work shows that the myopic sensing policy is not guaranteed to be optimal when the user's objective is to find at least one good channel.[3]

[2]In [7], the expected slot reward function is defined as $U(\mathcal{A}(t), \Omega(t)) = \sum_{i \in \mathcal{A}(t)} \omega_i(t)$.

[3]In [10], the expected slot reward function is defined as $U(\mathcal{A}(t), \Omega(t)) = 1 - \prod_{i \in \mathcal{A}(t)} (1 - \omega_i(t))$.

Given that such nuance on the reward function leads to totally contrary results, a natural while fundamentally important question arises:

How does the expected slot reward function $U(\mathcal{A}(t), \Omega(t))$ impact the optimality of the myopic sensing policy? Or more specifically, under what conditions on $U(\mathcal{A}(t), \Omega(t))$ is the myopic sensing policy guaranteed to be optimal?

2.3 Axioms

This section introduces a set of three axioms characterizing a family of generic and practically important functions, to which we refer as *regular* functions. The axioms developed in this section and the implied fundamental properties serve as a basis for the further analysis on the structure and the optimality of the myopic sensing policy in Sect. 2.4.

First, we state some structural properties of $\Gamma(\omega)$ and $\varphi(\omega)$ that are useful in the subsequent proofs.

Lemma 2.1 *For positively correlated channel,* i.e., $p_{01} < p_{11}$, *we have*

(i) $\Gamma(\omega)$ is monotonically increasing in ω
(ii) $p_{01} \le \Gamma(\omega) \le p_{11}, \forall\, 0 \le \omega \le 1$

Proof It follows from $\Gamma(\omega) = (p_{11} - p_{01})\omega + p_{01}$ straightforwardly. □

Lemma 2.2 *If* $0 \leqslant \mathcal{E} \leqslant \frac{(1-p_{11})p_{01}}{p_{11}(1-p_{01})}$ and $p_{01} < p_{11}$, then

(i) $\varphi(\omega)$ increases monotonically in ω with $\varphi(0) = 0$ and $\varphi(1) = 1$;

(ii) $\varphi(\omega) \leqslant p_{01}, \forall p_{01} \leqslant \omega \leqslant p_{11}$.

Proof Noticing that $\varphi(\omega) = \frac{\varepsilon\omega}{\varepsilon\omega+1-\omega}$, the lemma follows straightforwardly.

Throughout this section, for the convenience of presentation, we sort the elements of the believe vector $\Omega(t) = |\,\omega_1(t), \cdots, \omega_N(t)|$ for each slot t such that $\mathcal{A} = \{1, \cdots, k\}$ (i.e., the user senses channel 1 to channel k) and let $\Omega_{\mathcal{A}} \triangleq \{\omega_i : i \in \mathcal{A}\} = \{\omega_1, \cdots, \omega_k\}$[4].[4] The three axioms derived in the following characterize a generic function f defined on Ω_A. □

Axiom 1 (Symmetry) *A function* $f(\Omega_A) : [0, 1]^k \rightarrow \mathbb{R}$ *is symmetrical if* $\forall i, j \in \mathcal{A}$ *it holds that*

[4]For presentation simplicity, by slightly abusing the notations without introducing ambiguity, we drop the time slot index t.

$$f(\omega_1, \cdots, \omega_i, \cdots, \omega_j, \cdots, \omega_k) = f(\omega_1, \cdots, \omega_j, \cdots, \omega_i, \cdots, \omega_k) \tag{2.11}$$

Axiom 2 (Monotonicity) *A function* $f(\Omega_A) : [0, 1]^k \rightarrow \mathbb{R}$ *is monotonically increasing if it is monotonically increasing in each variable* ω_i, i.e., $\forall i \in \mathcal{A}$

$$\omega_i' > \omega_i \Rightarrow f(\omega_1, \cdots, \omega_i', \cdots, \omega_k) > f(\omega_1, \cdots, \omega_i, \cdots, \omega_k) \tag{2.12}$$

Axiom 3 (Decomposability) *A function* $f(\Omega_A) : [0, 1]^k \rightarrow \mathbb{R}$ *is decomposable if* $\forall i \in \mathcal{A}$ *it holds that*

$$\begin{aligned} f(\omega_1, \cdots, \omega_i, \cdots, \omega_k) = & \ \omega_i f(\omega_1, \cdots, 1, \cdots, \omega_k) \\ & + (1 - \omega_i) f(\omega_1, \cdots, 0, \cdots, \omega_k) \end{aligned} \tag{2.13}$$

Axioms 1 and 2 are intuitive. Axiom 3 on the decomposability states that $f(\Omega_A)$ can always be decomposed into two terms that replace ω_i by 0 and 1, respectively. The three axioms introduced in this section are consistent and non-redundant. Moreover, they can be used to characterize a family of generic functions, referred to as *regular* functions, defined as follows:

Definition 2.2 (Regular Function) A function is called regular if it satisfies all the three axioms.

The following definition studies the structure of the myopic sensing policy if the expected reward function is regular.

Proposition 2.1 (Structure of Myopic Sensing Policy) *Sort the elements of the belief vector in descending order such that* $\omega_1 \geq \cdots \geq \omega_N$, *if the expected slot reward function* $U(\bullet)$ *is regular, then the myopic sensing policy, where the user is allowed to sense k channels, consists of sensing the best k channels,* i.e., *channel* 1, . . ., *k*.

Remark 2.2 In case of tie, we sort the channels in tie in the descending order of $\omega_i(t + 1)$ calculated in (2.3). The argument is that larger $\omega_i(t + 1)$ leads to larger expected payoff in next slot $t + 1$. If the tie persists, the channels are sorted by indexes.

The developed three axioms characterize a set of generic functions widely used in practical applications. To see this, we give two examples to get more insight:

(i) The user gets one unit of reward for each channel that is sensed good and is indeed good. In this example, the expected reward function (for each slot) is $U(\mathcal{A}(t), \Omega(t)) = \sum_{i=1}^{k} [(1 - \mathcal{E})\omega_i(t)]$

(ii) The user gets one unit of reward if at least one channel is sensed good. In this example, the expected reward function is $U(\mathcal{A}(t), \Omega(t)) = 1 - \prod_{i=1}^{k} [1 - (1 - \mathcal{E})\,\omega_i(t)]$.

It is easy to verify that $U(\bullet)$ in both examples is regular since three axioms are satisfied.

2.4 Optimality of Myopic Sensing Policy under Imperfect Sensing

In this section we would establish closed-form conditions under which the myopic sensing policy, despite of its simple structure, achieves the system optimum under imperfect sensing. To this end, we set up by defining an auxiliary function and studying the structural properties of the auxiliary function, which serve as a basis in the study of the optimality of the myopic sensing policy. We then establish the main result on the optimality followed by the illustrating how the obtained result can be applied via two concrete application examples.

For the convenience of discussion, we first state some notations before presenting the analysis:

- The believe vector $\Omega(t)$ is sorted to $[\omega_1(t), \cdots, \omega_N(t)]$ each slot t such that $\mathcal{A}(t) = \{1, 2, \cdots, k\}$;
- $\mathcal{N}(m) \triangleq \{1, \cdots, m\} (m \leq N)$ denotes the first m channels in $\mathcal{N}$;
- Given $\mathcal{E} \subseteq \mathcal{M} \subseteq \mathcal{N}$,

$$\Pr(\mathcal{M}, \mathcal{E}) := \prod_{i \in \mathcal{E}} (1 - \mathcal{E}) \omega_i(t) \prod_{j \in \mathcal{M} \setminus \mathcal{E}} [1 - (1 - \mathcal{E}) \omega_j(t)] \tag{2.14}$$

where $\Pr(\mathcal{M}, \mathcal{E})$ denotes the expected probability that the channels in $\mathcal{E}$ are sensed to be in good state, while the channels in $\mathcal{M} \setminus \mathcal{E}$ are sensed in bad state, given that the channels in $\mathcal{M}$ are sensed;

- $\mathbf{P}_{11}^{\mathcal{E}}$ denotes the vector of length $|\mathcal{E}|$ with each element being p_{11};
- $\Phi(l, m) \triangleq [\Gamma(\omega_i(t)), l \leq i \leq m]$ where the components are sorted by channel index. $\Phi(l, m)$ characterizes the updated belief values of the channels between l and m if they are not sensed;
- Given $\mathcal{E} \subseteq \mathcal{M} \subseteq \mathcal{N}$,

$$\mathbf{Q}^{\mathcal{M}, \mathcal{E}} := [\Gamma(\varphi(\omega_i(t))) : i \in \mathcal{M} \setminus \mathcal{E}] \tag{2.15}$$

where the components are sorted by channel index. $\mathbf{Q}^{\mathcal{M}, \mathcal{E}}$ characterizes the updated belief values of the channels in $\mathcal{M} \setminus \mathcal{E}$ if they are sensed in the bad state;

$$\bar{\mathbf{Q}}^{\mathcal{M}, \mathcal{E}, l} := [\Gamma(\varphi(\omega_i(t))) : i \in \mathcal{M} \setminus \mathcal{E} \text{ and } i < l] \tag{2.16}$$

characterizes the updated belief values of the channels in $\mathcal{M}\backslash\mathcal{E}$ if they are sensed in the bad state with the channel index smaller than l;

$$\underline{\mathbf{Q}}^{\mathcal{M},\mathcal{E},l} := [\Gamma(\varphi(\omega_i(t))) : i \in \mathcal{M}\backslash\mathcal{E} \text{ and } i > l] \tag{2.17}$$

characterizes the updated belief values of the channels in $\mathcal{M} \setminus \mathcal{E}$ if they are sensed in the bad state with the channel index larger than l;

Let $\omega_{-i} := \{\omega_j : j \in \mathcal{A}, j \neq i\}$ and

$$\begin{cases} \Delta_{\max} := \max\limits_{i \in \mathcal{N}, \omega_{-i} \in [0,1]^{k-1}} \{U(1, \omega_{-i}) - U(0, \omega_{-i})\} \\ \Delta_{\min} := \min\limits_{i \in \mathcal{N}, \omega_{-i} \in [0,1]^{k-1}} \{U(1, \omega_{-i}) - U(0, \omega_{-i})\} \end{cases} \tag{2.18}$$

2.4.1 *Definition and Properties of Auxiliary Value Function*

In this subsection, inspired by the form of the value function $V_t(\Omega(t))$ and the analysis in [8], we first define the auxiliary value function with imperfect sensing and then derive several fundamental properties of the auxiliary value function, which are crucial in the study on the optimality of the myopic sensing policy.

Definition 2.3 (Auxiliary Value Function) The auxiliary value function, denoted as $W_t(\Omega(t))(t = 0, 1, \cdots, T)$, is recursively defined as follows:

$$W_T^{\widehat{\mathcal{A}}}(\Omega(T)) = U(\widehat{\mathcal{A}}(T), \Omega(T))$$

$$W_r^{\widehat{\mathcal{A}}}(\Omega(r)) = U(\widehat{\mathcal{A}}(r), \Omega(r)) + \underbrace{\beta \sum_{\mathcal{E} \subseteq \widehat{\mathcal{A}}(r)} \Pr(\widehat{\mathcal{A}}(r), \mathcal{E}) W_{r+1}^{\widehat{\mathcal{A}}}(\Omega_{\mathcal{E}}(r+1))}_{F(\widehat{\mathcal{A}}(r), \Omega(r))}$$

$$W_t^{\mathcal{A}}(\Omega(t)) = U(\mathcal{A}(t), \Omega(t)) + \underbrace{\beta \sum_{\mathcal{E} \subseteq \mathcal{A}(t)} \Pr(\mathcal{A}(t), \mathcal{E}) W_{t+1}^{\widehat{\mathcal{A}}}(\Omega_{\mathcal{E}}(t+1))}_{F(\mathcal{A}(t), \Omega(t))} \tag{2.19}$$

where $t < r < T$ and $\Omega_{\mathcal{E}}(t+1) := (\mathbf{P}_{11}^{\mathcal{E}}, \Phi(k+1, N), \mathbf{Q}^{\mathcal{A}(t),\mathcal{E}})$ denotes the belief vector generated by $\Omega(t)$ based on (2.3).

The above recursively defined auxiliary value function gives the expected cumulated reward of the following sensing policy: in slot t, sense the first k channels; if channel i is correctly sensed good, then put it on the top of the list to be sensed in next slot, otherwise drop it to the bottom of the list. Recall Lemmas 2.1 and 2.2, under the condition $0 \leqslant \mathcal{E} \leqslant \frac{(1-p_{11})p_{01}}{p_{11}(1-p_{01})}$, if the belief vector $\Omega(t)$ is ordered decreasingly in slot t, the above sensing policy is the myopic sensing policy with $W_t^{\mathcal{A}}(\Omega(t))$ being the total reward from slot t to T.

In the subsequent analysis, we prove some structural properties of the auxiliary value function.

Lemma 2.3 (Symmetry) *If the expected reward function $U(\bullet)$ is regular, the correspondent auxiliary value function $W_t^{\mathcal{A}}(\Omega(t))$ is symmetrical in any two channel $i, j \in \mathcal{A}(t)$ for all $t = 0, 1, \ldots, T$*, i. e.

$$W_t^{\mathcal{A}}(\omega_1, \cdots, \omega_i, \cdots, \omega_j, \cdots, \omega_N) = W_t^{\mathcal{A}}(\omega_1, \cdots, \omega_j, \cdots, \omega_i, \cdots, \omega_N), \quad \forall i, j \le k \tag{2.20}$$

Proof The lemma can be easily shown by backward induction noticing that at slot t, $(\cdots, \omega_i, \cdots, \omega_j, \cdots)$ and $(\cdots, \omega_j, \cdots, \omega_i, \cdots)$ generate the same belief vector $\Omega_{\mathcal{E}}(t+1)$ for any $\mathcal{E}$. □

Lemma 2.4 (Decomposability) *If the expected reward function $U(\bullet)$ is regular, then the correspondent auxiliary value function $W_t^{\mathcal{A}}(\Omega(t))$ is decomposable for all $t = 0, 1, \cdots, T$*, i. e.

$$\begin{aligned} W_t^{\mathcal{A}}(\omega_1, \cdots, \omega_i, \cdots, \omega_N) &= \omega_i W_t^{\mathcal{A}}(\omega_1, \cdots, 1, \cdots, \omega_N) \\ &+ (1 - \omega_i) W_t^{\mathcal{A}}(\omega_1, \cdots, 0, \cdots, \omega_N), \quad \forall i \in \mathcal{N} \end{aligned} \tag{2.21}$$

Proof The proof is given in the Appendix A.

Lemma 2.4 can be applied one step further to prove the following corollary. □

Corollary 2.1 *If the expected reward function $U(\bullet)$ is regular, then for any $l; m \in \mathcal{N}$ holds that for $t = 0, 1, \cdots, T$*

$$\begin{aligned} &W_t^{\mathcal{A}}(\omega_1, \cdots, \omega_l, \cdots, \omega_m, \cdots, \omega_N) - W_t^{\mathcal{A}}(\omega_1, \cdots, \omega_m, \cdots, \omega_l, \cdots, \omega_N) \\ &\quad = (\omega_l - \omega_m) \\ &\qquad \times \left[W_t^{\mathcal{A}}(\omega_1, \cdots, 1, \cdots, 0, \cdots, \omega_N) - W_t^{\mathcal{A}}(\omega_1, \cdots, 0, \cdots, 1, \cdots, \omega_N)\right] \end{aligned} \tag{2.22}$$

Lemma 2.5 (Monotonicity) *If the expected reward function U is regular, the correspondent auxiliary value function $W_t^{\mathcal{A}}(\Omega)$ is monotonously non-decreasing in ω_l, $\forall l \in \mathcal{N}$*, i.e.,

$$\omega_l' \ge \omega_l \Rightarrow W_t^{\mathcal{A}}(\omega_1, \cdots, \omega_l', \cdots, \omega_N) \ge W_t^{\mathcal{A}}(\omega_1, \cdots, \omega_l, \cdots, \omega_N) \tag{2.23}$$

Proof The proof is given in the Appendix A. □

2.4.2 *Optimality of Myopic Sensing: Positively Correlated Channels*

In this section, we study the optimality of the myopic sensing policy under imperfect sensing. We start by showing the following important auxiliary lemmas (Lemmas 2.6 and 2.7) and then establish the sufficient condition under which the optimality of the myopic sensing policy is guaranteed.

Lemma 2.6 *Given*

(i) $\mathcal{E} < \frac{p_{01}(1-p_{11})}{P_{11}(1-p_{01})}$',

(ii) $\beta \leq \frac{\Delta_{\min}/\Delta_{\max}}{(1-\mathcal{E})(1-p_{01})+\frac{\mathcal{E}(p_{11}-p_{01})}{1-(1-\mathcal{E})(p_{11}-p_{01})}}$,

(iii) $U(\cdot)$is regular,

if $p_{11} \geq \omega_l \geq \omega_m \geq p_{01}$ *where* $l < m$*, then it holds that for* $t = 0, 1, \cdots, T$

$$W_t^{\mathcal{A}}(\omega_1, \cdots, \omega_l, \cdots, \omega_m, \cdots, \omega_N) \geq W_t^{\mathcal{A}}(\omega_1, \cdots, \omega_m, \cdots, \omega_l, \cdots, \omega_N) \tag{2.24}$$

Lemma 2.7 *Given*

(i) $\mathcal{E} < \frac{p_{01}(1-p_{11})}{P_{11}(1-p_{01})}$,

(ii) $\beta \leq \frac{\Delta_{\min}/\Delta_{\max}}{(1-\mathcal{E})(1-p_{01})+\frac{\mathcal{E}(p_{11}-p_{01})}{1-(1-\mathcal{E})(p_{11}-p_{01})}}$,

(iii) $U(\cdot)$is regular,

If $p_{11} \geq \omega_1 \geq \cdots \geq \omega_N \geq p_{01}$,for any $0 \leq t \leq T$, it holds that

$$W_t^{\mathcal{A}}(\omega_1, \omega_2, \cdots, \omega_{N-1}, \omega_N) - W_t^{\mathcal{A}}(\omega_N, \omega_1, \cdots, \omega_{N-1}) \leq (1-\omega_N)\Delta_{\max},$$

$$W_t^{\mathcal{A}}(\omega_1, \omega_2, \cdots, \omega_{N-1}, \omega_N) - W_t^{\mathcal{A}}(\omega_N, \omega_2, \cdots, \omega_{N-1}, \omega_1)$$
$$\leq (p_{11} - p_{01})\Delta_{\max} \frac{1 - [\beta(1-\mathcal{E})(p_{11}-p_{01})]^{T-t+1}}{1 - \beta(1-\mathcal{E})(p_{11}-p_{01})}$$

Lemma 2.6 states that by swapping two elements in Q with the former larger than the latter, it cannot increase the total expected reward. Lemma 2.7, on the other hand, gives the upper bound on the difference of the total reward of the two swapping operations, swapping ω_N and $\omega_k (k = N - 1, \cdots, 1)$ and swapping «1 and *mN*, respectively. For clarity of presentation, the detailed proofs of the two lemmas are deferred to the Appendix A.

Theorem 2.1 *If* $p_{01} \leqslant \omega_i(0) \leqslant p_{11}$ *for any* $i(1 \leqslant i \leqslant N)$*, the myopic sensing policy is optimal if the following conditions hold*

(i) $\mathcal{E} < \frac{p_{01}(1-p_{11})}{p_{11}(1-p_{01})}$,

(ii) $\beta \leq \frac{\Delta_{\min}/\Delta_{\max}}{(1-\mathcal{E})(1-p_{01})+\frac{\mathcal{E}(p_{11}-p_{01})}{1-(1-\mathcal{E})(p_{11}-p_{01})}}$,

(iii) $U(\cdot)$is regular.

Proof It suffices to show that for $t = 0, 1, \cdots, T$, by sorting $\Omega(t)$ in decreasing order such that $\omega_1 \geq \cdots \geq \omega_N$, it holds that $W_t^{\mathcal{A}}(\omega_1, \cdots, \omega_N) \geq W_t^{\mathcal{A}}(\omega_{i_1}, \cdots, \omega_{i_N})$, where $(\omega_{i_1}, \cdots, \omega_{i_N})$ is any permutation of $(1, \cdots, N)$.

We prove the above inequality by contradiction. Assume, by contradiction, the maximum of $W_t(\cdot)$ is achieved at $\left(\omega_{i_1^*}, \cdots, \omega_{i_N^*}\right) \neq (\omega_1, \cdots, \omega_N)$, i.e. ,

$$W_t^{\mathcal{A}}\left(\omega_{i_1^*}, \cdots, \omega_{i_N^*}\right) > W_t^{\mathcal{A}}(\omega_1, \cdots, \omega_N) \tag{2.25}$$

However, we run a bubble sort algorithm on $\left(\omega_{i_1^*}, \cdots, \omega_{i_N^*}\right)$ by repeatedly stepping through it, comparing each pair of adjacent element $\omega_{i_l^*}$ and $\omega_{i_{l+1}^*}$ and swapping them if $\omega_{i_l^*} < \omega_{i_l^*+1}$ When the algorithm terminates, the elements of the channel belief vector are sorted decreasingly; that is to say, the belief vector becomes $(\omega_1, \cdots, \omega_N)$ finally. By applying Lemma 2.6 at each swapping, we obtain $W_t^{\mathcal{A}}\left(\omega_{i_1^*}, \cdots, \omega_{i_N^*}\right) \leq W_t^{\mathcal{A}}(\omega_1, \cdots, \omega_N)$, which contradicts (2.25). Theorem 2.1 is thus proven.

As noted in [12], when the initial belief ω_i is set to $\frac{p_{01}}{p_{01}+1-p_{11}}$ as the popular case in practical systems, it can be checked that $p_{01} \leqslant \omega_i(0) \leqslant p_{11}$ holds. Moreover, even the initial belief does not fall in $[p_{01}, p_{11}]$, all the belief values are bounded in the interval from the second slot following Lemma 2.1. Hence our results can be extended by treating the first slot separately from the future slots. □

2.4.3 Discussion

In this subsection, we illustrate the application of the obtained result in two concrete scenarios and compare our work with the existing results.

Problem 2.1 Consider the channel access problem in which a user is limited to sensing k channels and gets one unit of reward if the sensed channel is in the good state, i.e., the utility function can be formulated as $U(\mathcal{A}(t), \Omega_A) = (1 - \mathcal{E}) \sum_{i\in\mathcal{A}(t)} \omega_i$.

The optimality of the myopic sensing policy for this model was studied in [12] for a subset of scenarios where $k = 1$ and $N = 2$. We now study the generic case with $k, N \geq 2$. To that end, we have $\Delta_{\min} = \Delta_{\max} = 1 - \varepsilon$, and then can verify that when $\mathcal{E} < \frac{p_{01}(1-p_{11})}{p_{11}(1-p_{10})}$, it holds that $\frac{\Delta_{\min}}{\Delta_{\max}\left[(1-\mathcal{E})(1-p_{01})+\frac{\mathcal{E}(p_{11}-p_{01})}{1-(1-\mathcal{E})(p_{11}-p_{01})}\right]} > 1$.. Therefore, when the conditions (i) and (ii) hold, the myopic sensing policy is optimal for $0 \leqslant \beta \leqslant 1$ by Theorem 2.1. This result in generic cases significantly extends the optimality [12] in which the optimality of the myopic policy is proved for the case of two channels and only conjectured for general cases.

Problem 2.2 Consider the channel access scenario where a user can sense k channels but can only choose one of them to transmit its packets. Under this model, the user wants to maximize its expected throughput. More specifically, the slot utility function$U(\mathcal{A}(t), \Omega(t)) = 1 - \Pi_{i\in\mathcal{A}(t)}[1 - (1 - \mathcal{E})\omega_i(t)]$, which is regular.

In this context, $\Delta_{\max} = (1 - \mathcal{E})^{k-1} p_{11}^{k-1}$ and $\Delta_{\min} = (1 - \mathcal{E})^{k-1} p_{01}^{k-1}$. The third condition $\beta \leq \frac{p_{01}^{k-1}}{p_{11}^{k-1}\left[(1-\mathcal{E})(1-p_{01})+\frac{\mathcal{E}(p_{11}-p_{01})}{1-(1-\mathcal{E})(p_{11}-p_{01})}\right]}$ Particularly, when $\varepsilon = 0$, $\beta \leq \frac{p_{01}^{k-1}}{p_{11}^{k-1}(1-p_{01})}$. It can be noted that even though there is no sensing error, the myopic policy is not guaranteed to be optimal.

Since the conditions in Theorem 2.1 is sufficient, it is insightful to see how tight the condition is, especially the third condition. To this end, we provide an example in which the third condition is only slightly violated and the myopic sensing policy is not optimal.

Example 2.1 $T = 2$, $\Omega(0) = [0.5814, 0.41, 0.40, 0.33, 0.32, 0.31]$, $p_{11} = 0.5815$ and $p_{01} = 0.3$, $k = 3$, $\beta = 0.99992$, $\mathcal{E} = 0.25 < \frac{p_{01}(1-p_{11})}{p_{11}(1-p_{01})} = 0.40$, $U(\cdot) = 1 - \prod_{i\in\mathcal{A}}(1 - \omega_i)$ which is regular.

In this example, it can be checked that sensing channels 1; 2; 4 yields more payoff than the myopic sensing policy since

$$\frac{\Delta_{\min}}{\Delta_{\max}\left[(1 - \mathcal{E})(1 - p_{01}) + \frac{\mathcal{E}(p_{11}-p_{01})}{1-(1-\mathcal{E})(p_{11}-p_{01})}\right]} = 0.9917 < 0.99992 = \beta.$$

This example evidences that the condition in Theorem 2.1 is quite tight such that a slight violation can lead to the non-optimality of the myopic sensing policy.

2.5 Optimality Extension to Negatively Correlated Channels

In this section, we study the optimality of the proposed myopic policy for negatively correlated homogenous channels, i.e., $p_{11} < p_{01}$.

In [9], the authors showed by a counterexample that the myopic policy is not optimal for negatively correlated homogenous channels. However, we would prove that the myopic policy is optimal only by imposing a weak condition on the initial belief vector $\Omega(0)$, i. e. , $\forall\, i$, $p_{11} \leqslant \omega_i(0) \leqslant p_{01}$. In fact, the weak condition will be automatically satisfied from the second slot since the belief value would enter into $[p_{11}, p_{01}]$ by the operator $\Gamma(\cdot)$.

Lemma 2.8 *For negatively correlated channel, i.e., $p_{01} > p_{11}$, we have*

(i) $\Gamma(\omega)$ is monotonically decreasing in ω;
(ii) $p_{11} \leq \Gamma(\omega) \leq p_{01}$, $\forall\, 0 \leq \omega \leq 1$.

Lemma 2.9 If $0 \leqslant \mathcal{E} \leqslant \frac{(1-p_{01})p_{11}}{(1-p_{11})p_{01}}$ and $p_{11} < p_{01}$, then

(i) *$\varphi(\omega)$ increases monotonically in ω with $\varphi(0) = 0$ and $\varphi(1) = 1$;*
(ii) $\varphi(\omega) \leqslant p_{11}$, $\forall\, p_{11} \leqslant \omega \leqslant p_{01}$

Through analyzing the queue structure of the myopic policy [5], we find out that the proof for positively correlated homogenous model can be slightly modified to fit into the negatively correlated homogenous model. Hence, we first give the following structure of belief vector, and then points out the nuance in the proof of optimality of myopic policy.

Theorem 2.2 (Structure of Myopic Policy) *If $\mathcal{E} \leqslant \frac{p_{11}(1-p_{01})}{p_{01}(1-p_{11})}$, we have the following channel order rules at the end of each slot.*

(i) The initial channel ordering $Q(0)$ is determined by the initial belief vector:

$$\omega_{\sigma_1}(0) \geqslant \cdots \geqslant \omega_{\sigma_N}(0) \Rightarrow \mathbf{Q}(0) = (\sigma_1, \sigma_2, \cdots, \sigma_N)$$

(ii) The channels sensed to be in bad state will move to the end of the queue while the channels sensed to be in good state will stay at the head of the queue, and the order of other channels will be reversed.

Proof Assume $\mathbf{Q}(t) = (\sigma_1, \cdots, \sigma_N)$ at slot t, we thus have $p_{01} \geqslant \omega_{\sigma_1}(t) \geqslant \cdots \geqslant \omega_{\sigma_N}(t) \geqslant p_{11}$. If channel σ_1 is sensed to be in good state, then $\omega_{\sigma_1}(t+1) = \Gamma(1) \leqslant \Gamma(\omega_{\sigma_2}(t)) \leqslant \cdots \leqslant \Gamma(\omega_{\sigma_N}(t))$ by Lemma 2.8, and thus $\mathbf{Q}(t+1) = (\sigma_N, \cdots, \sigma_1)$ according to the descending order of «.If channel σ_1 is sensed to be in bad state, then $\omega_{\sigma_1}(t+1) = \Gamma(\varphi(\omega_{\sigma_1}(t))) \geqslant \Gamma(p_{11}) \geqslant \Gamma(\omega_{\sigma_N}(t)) \geqslant \cdots \geqslant \Gamma(\omega_{\sigma_2}(t))$, and further $\mathbf{Q}(t+1) = (\sigma_1, \sigma_N, \cdots, \sigma_2)$. □

Table 2.2 Structure of myopic policy with $\mathbf{Q}(t) = (\sigma_1, \cdots, \sigma_N)$

Sensed state of channel σ_1	Positively correlated	Negatively correlated
Good	$\mathbf{Q}(t+1) = (\sigma_1, \sigma_2, \cdots, \sigma_N)$	$\mathbf{Q}(t+1) = (\sigma_N, \cdots, \sigma_2, \sigma_1)$
Bad	$\mathbf{Q}(t+1) = (\sigma_2, \cdots, \sigma_N, \sigma_1)$	$\mathbf{Q}(t+1) = (\sigma_1, \sigma_N, \cdots, \sigma_2)$

Remark 2.3 Assume $\mathbf{Q}(t) = (\sigma_1, \cdots, \sigma_N,)$ at slot t where $\omega_{\sigma_1}(t) \geqslant \cdots \geqslant \omega_{\sigma_N}(t)$. When channel σ_1 is sensed to in good state or bad state, respectively, the structure of $\mathbf{Q}(t+1)$ is stated in the following table. Meanwhile, $\mathbf{Q}(t+1)$ for positively correlated homogeneous channels is also listed for the purpose of comparison. As shown in Table 2.2, $\mathbf{Q}(t+1)$shows the reverse order in two cases. It is the reverse order which preserves two kinds of exchange operation in Lemmas 2.6 and 2.7. Thus, Lemmas 2.6 and 2.7 still hold by exchanging p_{11} and p_{01}.

Following the similar induction for positively correlated channels, we have the following lemmas and theorem without proof.

Lemma 2.10 *Given*

(i) $\mathcal{E} < \frac{p_{11}(1-p_{01})}{p_{01}(1-p_{11})}$,
(ii) $\beta \leq \frac{\Delta_{\min}/\Delta_{\max}}{(1-\mathcal{E})(1-p_{11}) + \frac{\mathcal{E}(p_{01}-p_{11})}{1-(1-\mathcal{E})(p_{01}-p_{11})}}$,
(iii) $U(\cdot)$ is regular,

if $p_{01} \geq \omega_l \geq \omega_m \geq p_{11}$ *where* $l < m$*, then it holds that for* $t = 0, 1, \cdots, T$

$$W_t^{\mathcal{A}}(\omega_1, \cdots, \omega_l, \cdots, \omega_m, \cdots, \omega_N) \geq W_t^{\mathcal{A}}(\omega_1, \cdots, \omega_m, \cdots, \omega_l, \cdots, \omega_N) \tag{2.26}$$

Lemma 2.11 *Given*

(i) $\mathcal{E} < \frac{p_{11}(1-p_{01})}{p_{01}(1-p_{11})}$,
(ii) $\beta \leq \frac{\Delta_{\min}/\Delta_{\max}}{(1-\mathcal{E})(1-p_{11}) + \frac{\mathcal{E}(p_{01}-p_{11})}{1-(1-\mathcal{E})(p_{01}-p_{11})}}$,
(iii) $U(\cdot)$ is regular.

If $p_{01} \geq \omega_1 \geq \cdots \geq \omega_N \geq p_{11}$, *for any* $0 \leq t \leq T$*, it holds that*

$$W_t^{\mathcal{A}}(\omega_1, \omega_2, \cdots, \omega_{N-1}, \omega_N) - W_t^{\mathcal{A}}(\omega_N, \omega_1, \cdots, \omega_{N-1}) \leq (1-\omega_N)\Delta_{\max},$$

$$W_t^{\mathcal{A}}(\omega_1, \omega_2, \cdots, \omega_{N-1}, \omega_N) - W_t^{\mathcal{A}}(\omega_N, \omega_2, \cdots, \omega_{N-1}, \omega_1)$$
$$\leq (p_{01} - p_{11})\Delta_{\max} \frac{1 - [\beta(1-\mathcal{E})(p_{01} - p_{11})]^{T-t+1}}{1 - \beta(1-\mathcal{E})(p_{01} - p_{11})}.$$

Theorem 2.3 *If $p_{11} \leqslant \omega_i(0) \leqslant p_{01}$ for $1 \leqslant i \leqslant N$, the myopic policy is optimal if*

(i) $\mathcal{E} < \frac{p_{11}(1-p_{01})}{p_{01}(1-p_{11})}$,

(ii) $\beta \leq \frac{\Delta_{\min}/\Delta_{\max}}{(1-\mathcal{E})(1-p_{11})+\frac{\mathcal{E}(p_{01}-p_{11})}{1-(1-\mathcal{E})(p_{01}-p_{11})}}$,

(iii) $U(\cdot)$ is regular.

Remark 2.4 Theorem 2.3 gives the sufficient conditions to justify the optimality of the myopic policy, i.e., probing those best channels, for the negatively correlated homogeneous channels. More importantly, this theorem contradicts the intuition that the myopic policy is not optimal for negatively correlated case.

2.6 Summary

In this chapter, we have investigated the optimality of the myopic policy in the context of opportunistic access with imperfect channel sensing for two-state Markov channel, and derived closed-form conditions under which the myopic sensing policy is ensured to be optimal for homogeneous two-state Markov channels. Due to the generic RMAB formulation of the problem, the obtained results and the analysis methodology in this chapter are widely applicable in a wide range of engineering domains.

Appendix

Proof of Lemma 2.4

We proceed the proof by backward induction. Firstly, it is easy to verify that the lemma holds for slot T.

Assume that the lemma holds from slots $t+1, \cdots, T-1$, we now prove that it holds for slot t by the following two different cases.

Case 1: channel l is not sensed in slot t, i.e., $l \geq k+1$. Let $\mathcal{A}(t) = \mathcal{M} \triangleq \mathcal{N}(k) = \{1, \cdots, k\}$, $\omega_l = 0$ and 1, respectively, we have

$$W_t^{\mathcal{A}}(\omega_1, \cdots, \omega_l, \cdots, \omega_n) = U(\omega_1, \cdots, \omega_k) + \beta \sum_{\mathcal{E} \subseteq \mathcal{M}} \Pr(\mathcal{M}, \mathcal{E}) W_{t+1}^{\widehat{\mathcal{A}}}(\Omega_l^{\mathcal{E}}(t+1)),$$

$$W_t^{\mathcal{A}}(\omega_1, \cdots, 0, \cdots, \omega_n) = U(\omega_1, \cdots, \omega_k) + \beta \sum_{\mathcal{E} \subseteq \mathcal{M}} \Pr(\mathcal{M}, \mathcal{E}) W_{t+1}^{\widehat{\mathcal{A}}}(\Omega_{l,0}^{\mathcal{E}}(t+1)),$$

$$W_t^{\mathcal{A}}(\omega_1, \cdots, 1, \cdots, \omega_n) = U(\omega_1, \cdots, \omega_k) + \beta \sum_{\mathcal{E} \subseteq \mathcal{M}} \Pr(\mathcal{M}, \mathcal{E}) \widehat{W_{t+1}^{\mathcal{A}}}(\Omega_{l,1}^{\mathcal{E}}(t+1)),$$

where

$$\Omega_l^{\mathcal{E}}(t+1) = (\mathbf{P}_{11}^{\mathcal{E}}, \Phi(k+1, l-1), \Gamma(\omega_l), \Phi(l+1, N), \mathbf{Q}^{\mathcal{M},\mathcal{E}}),$$
$$\Omega_{l,0}^{\mathcal{E}}(t+1) = (\mathbf{P}_{11}^{\mathcal{E}}, \Phi(k+1, l-1), p_{01}, \quad \Phi(l+1, N), \mathbf{Q}^{\mathcal{M},\mathcal{E}}),$$
$$\Omega_{l.1}^{\mathcal{E}}(t+1) = (\mathbf{P}_{11}^{\mathcal{E}}, \Phi(k+1, l-1), p_{11}, \quad \Phi(l+1, N), \mathbf{Q}^{\mathcal{M},\mathcal{E}}).$$

To prove the lemma in this case, it is sufficient to prove

$$\begin{aligned} &\widehat{W_{t+1}^{\mathcal{A}}}(\Omega_l^{\mathcal{E}}(t+1)) \\ &= (1-\omega_l)\widehat{W_{t+1}^{\mathcal{A}}}(\Omega_{l,0}^{\mathcal{E}}(t+1)) + \omega_l \widehat{W_{t+1}^{\mathcal{A}}}(\Omega_{l,1}^{\mathcal{E}}(t+1)) \end{aligned} \tag{2.27}$$

According to induction hypothesis, we have

$$\begin{aligned} &\widehat{W_{t+1}^{\mathcal{A}}}(\Omega_l^{\mathcal{E}}(t+1)) \\ &= \Gamma(\omega_l) \cdot \widehat{W_{t+1}^{\mathcal{A}}}(\mathbf{P}_{11}^{\mathcal{E}}, \Phi(k+1, l-1), 1, \Phi(l+1, N), \mathbf{Q}^{\mathcal{M},\mathcal{E}}) \\ &+ (1-\Gamma(\omega_l)) \cdot \widehat{W_{t+1}^{\mathcal{A}}}(\mathbf{P}_{11}^{\mathcal{E}}, \Phi(k+1, l-1), 0, \Phi(l+1, N), \mathbf{Q}^{\mathcal{M},\mathcal{E}}) \end{aligned} \tag{2.28}$$

$$\begin{aligned} &\widehat{W_{t+1}^{\mathcal{A}}}(\Omega_{l,0}^{\mathcal{E}}(t+1)) \\ &= p_{01} \cdot \widehat{W_{t+1}^{\mathcal{A}}}(\mathbf{P}_{11}^{\mathcal{E}}, \Phi(k+1, l-1), 1, \Phi(l+1, N), \mathbf{Q}^{\mathcal{M},\mathcal{E}}) \\ &+ (1-p_{01}) \cdot \widehat{W_{t+1}^{\mathcal{A}}}(\mathbf{P}_{11}^{\mathcal{E}}, \Phi(k+1, l-1), 0, \Phi(l+1, N), \mathbf{Q}^{\mathcal{M},\mathcal{E}}) \end{aligned} \tag{2.29}$$

$$\begin{aligned} \widehat{W_{t+1}^{\mathcal{A}}}(\Omega_{l,1}^{\mathcal{E}}(t+1)) &= p_{11} \cdot \widehat{W_{t+1}^{\mathcal{A}}}(\mathbf{P}_{11}^{\mathcal{E}}, \Phi(k+1, l-1), 1, \Phi(l+1, N), \mathbf{Q}^{\mathcal{M},\mathcal{E}}) \\ &+ (1-p_{11}) \cdot \widehat{W_{t+1}^{\mathcal{A}}}(\mathbf{P}_{11}^{\mathcal{E}}, \Phi(k+1, l-1), 0, \Phi(l+1, N), \mathbf{Q}^{\mathcal{M},\mathcal{E}}) \end{aligned} \tag{2.30}$$

Combing (2.28), (2.29) with (2.30), we have (2.27).

Case 2: channel l is sensed in slot t, i.e., $l < k$. Let $\mathcal{M} \triangleq \mathcal{N}(k) \backslash \{l\} = \{1, \cdots, l-1, l+1, \cdots, k\}$, we have according to (2.19)

$$
\begin{aligned}
&W_t^{\mathcal{A}}(\Omega(t)) \\
&= U(\omega_1, \cdots, \omega_l, \cdots, \omega_k) \\
&+ \beta(1-\mathcal{E})\omega_l \sum_{\mathcal{E} \subseteq \mathcal{M}} \Pr(\mathcal{M}, \mathcal{E}) \widehat{W_{t+1}^{\mathcal{A}}}(\mathbf{P}_{11}^{\mathcal{E}}, p_{11}, \Phi(k+1, N), \bar{\mathbf{Q}}^{\mathcal{M},\mathcal{E},l}, \underline{\mathbf{Q}}^{\mathcal{M},\mathcal{E},l}) \\
&+ \beta[1-(1-\mathcal{E})\omega_l] \sum_{\mathcal{E} \subseteq \mathcal{M}} \Pr(\mathcal{M}, \mathcal{E}) \widehat{W_{t+1}^{\mathcal{A}}}(\mathbf{P}_{11}^{\mathcal{E}}, \Phi(k+1, N), \bar{\mathbf{Q}}^{\mathcal{M},\mathcal{E},l}, \Gamma(\varphi(\omega_l)), \underline{\mathbf{Q}}^{\mathcal{M},\mathcal{E},l})
\end{aligned} \tag{2.31}
$$

For (2.31), letting $\omega_l = 0$, we have

$$
\begin{aligned}
&W_t^{\mathcal{A}}(\omega_1, \cdots, 0, \cdots, \omega_N) \\
&= U(\omega_1, \cdots, 0, \cdots, \omega_k) \\
&\quad + \beta \sum_{\mathcal{E} \subseteq \mathcal{M}} \Pr(\mathcal{M}, \mathcal{E}) \widehat{W_{t+1}^{\mathcal{A}}}(\mathbf{P}_{11}^{\mathcal{E}}, \Phi(k+1, N), \bar{\mathbf{Q}}^{\mathcal{M},\mathcal{E},l}, p_{01}, \underline{\mathbf{Q}}^{\mathcal{H},\mathcal{E},l})
\end{aligned} \tag{2.32}
$$

For (2.31), letting $\omega_l = 1$, we have

$$
\begin{aligned}
&W_t^{\mathcal{A}}(\omega_1, \cdots, 1, \cdots, \omega_N) \\
&= U(\omega_1, \cdots, 1, \cdots, \omega_k) \\
&\quad + \beta(1-\mathcal{E}) \sum_{\mathcal{E} \subseteq \mathcal{M}} \Pr(\mathcal{M}, \mathcal{E}) \widehat{W_{t+1}^{\mathcal{A}}}(\mathbf{P}_{11}^{\mathcal{E}}, p_{11}, \Phi(k+1, N), \bar{\mathbf{Q}}^{\mathcal{M},\mathcal{E},l}, \underline{\mathbf{Q}}^{\mathcal{M},\mathcal{E},l}) \\
&\quad + \beta\mathcal{E} \sum_{\mathcal{E} \subseteq \mathcal{M}} \Pr(\mathcal{M}, \mathcal{E}) \widehat{W_{t+1}^{\mathcal{A}}}(\mathbf{P}_{11}^{\mathcal{E}}, \Phi(k+1, N), \bar{\mathbf{Q}}^{\mathcal{M},\mathcal{E},l}, p_{11}, \underline{\mathbf{Q}}^{\mathcal{M},\mathcal{E},l})
\end{aligned} \tag{2.33}
$$

To prove the lemma for this case (channel l is not sensed in slot t), based on (2.31), (2.32), and (2.33), we only need to show

$$
\begin{aligned}
&[1-(1-\mathcal{E})\omega_l] \widehat{W_{t+1}^{\mathcal{A}}}(\mathbf{P}_{11}^{\mathcal{E}}, \Phi(k+1, N), \bar{\mathbf{Q}}^{\mathcal{M},\mathcal{E},1}, \Gamma(\varphi(\omega_l)), \underline{\mathbf{Q}}^{\mathcal{M},\mathcal{E},\mathbf{l}}) \\
&\quad = (1-\omega_l) \widehat{W_{t+1}^{\mathcal{A}}}(\mathbf{P}_{11}^{\mathcal{E}}, \Phi(k+1, N), \bar{\mathbf{Q}}^{\mathcal{M},\mathcal{E},l}, p_{01}, \underline{\mathbf{Q}}^{\mathcal{M},\mathcal{E},l}) \\
&\qquad + \mathcal{E}\omega_l \widehat{W_{t+1}^{\mathcal{A}}}(\mathbf{P}_{11}^{\mathcal{E}}, \Phi(k+1, N), \bar{\mathbf{Q}}^{\mathcal{M},\mathcal{E},l}, p_{11}, \underline{\mathbf{Q}}^{\mathcal{M},\mathcal{E},l})
\end{aligned} \tag{2.34}
$$

According to induction hypothesis, we have

$$
\widehat{W_{t+1}^{\mathcal{A}}}(\mathbf{P}_{11}^{\mathcal{E}}, \Phi(k+1, N), \bar{\mathbf{Q}}^{\mathcal{M},\mathcal{E},l}, \Gamma(\varphi(\omega_l)), \underline{\mathbf{Q}}^{\mathcal{M},\mathcal{E},l})
$$

$$= \Gamma(\varphi(\omega_l))\widehat{W^{\mathcal{A}}_{t+1}}(\mathbf{P}^{\mathcal{E}}_{11}, \Phi(k+1,N), \bar{\mathbf{Q}}^{\mathcal{M},\mathcal{E},1}, 1, \underline{\mathbf{Q}}^{\mathcal{M},\mathcal{E},l})$$
$$+ (1 - \Gamma(\varphi(\omega_l)))\widehat{W^{\mathcal{A}}_{t+1}}(\mathbf{P}^{\mathcal{E}}_{11}, \Phi(k+1,N), \bar{\mathbf{Q}}^{\mathcal{M},\mathcal{E},l}, 0, \underline{\mathbf{Q}}^{\mathcal{M},\mathcal{E},l}) \quad (2.35)$$

$$\widehat{W^{\mathcal{A}}_{t+1}}(\mathbf{P}^{\mathcal{E}}_{11}, \Phi(k+1,N), \bar{\mathbf{Q}}^{\mathcal{M},\mathcal{E},l}, p_{01}, \underline{\mathbf{Q}}^{\mathcal{M},\mathcal{E},l})$$
$$= p_{01}\widehat{W^{A}_{t+1}}(\mathbf{P}^{\mathcal{E}}_{11}, \Phi(k+1,N), \bar{\mathbf{Q}}^{\mathcal{M},\mathcal{E},l}, 1, \underline{\mathbf{Q}}^{\mathcal{M},\mathcal{E},l})$$
$$+ (1 - p_{01})\widehat{W^{\mathcal{A}}_{t+1}}(\mathbf{P}^{\mathcal{E}}_{11}, \Phi(k+1,N), \bar{\mathbf{Q}}^{\mathcal{M},\mathcal{E},l}, 0, \underline{\mathbf{Q}}^{\mathcal{M},\mathcal{E},l}) \quad (2.36)$$

$$\widehat{W^{\mathcal{A}}_{t+1}}(\mathbf{P}^{\mathcal{E}}_{11}, \Phi(k+1,N), \bar{\mathbf{Q}}^{\mathcal{M},\mathcal{E},l}, p_{11}, \underline{\mathbf{Q}}^{\mathcal{M},\mathcal{E},l})$$
$$= p_{11}\widehat{W^{\mathcal{A}}_{t+1}}(\mathbf{P}^{\mathcal{E}}_{11}, \Phi(k+1,N), \bar{\mathbf{Q}}^{\mathcal{M},\mathcal{E},l}, 1, \underline{\mathbf{Q}}^{\mathcal{M},\mathcal{E},l})$$
$$+ (1 - p_{11})\widehat{W^{\mathcal{A}}_{t+1}}(\mathbf{P}^{\mathcal{E}}_{11}, \Phi(k+1,N), \bar{\mathbf{Q}}^{\mathcal{M},\mathcal{E},l}, 0, \underline{\mathbf{Q}}^{\mathcal{M},\mathcal{E},l}) \quad (2.37)$$

Combing (2.35), (2.36), and (2.37), we have (2.34).

Combing the above analysis in two cases, we thus prove Lemma 2.4.

Proof of Lemma 2.5

We proceed the proof by backward induction. Firstly, we can easily show that the lemma holds for slot T. Assume that the lemma holds from slots $t+1, \cdots, T-1$, we now prove that it also holds for slot t by distinguishing the following two cases.

Case 1: channel l is not sensed in slot t, i.e., $l \geq k+1$. In this case, the immediate reward is unrelated to ω_l and ω'_l. Moreover, let $\Omega(t+1)$ and $\Omega'(t+1)$ denote the belief vector generated by $\Omega(t) = (\omega_1, \cdots, \omega_l, \cdots, \omega_N)$ and $\Omega'(t) = (\omega_1, \cdots, \omega'_l, \cdots, \omega_N)$, respectively, it can be noticed that $\Omega(t+1)$ and $\Omega'(t+1)$ differ in only one element: $\omega'_l(t+1) \geq \omega_l(t+1)$. By induction, it holds that $\widehat{W^{\mathcal{A}}_{t+1}}(\Omega'(t+1)) \geq \widehat{W^{\mathcal{A}}_{t+1}}(\Omega(t+1))$. Noticing (2.19), it follows that $W^{\mathcal{A}}_t(\Omega'(t)) \geq W^{\mathcal{A}}_t(\Omega(t))$.

Case 2: channel l is sensed in slot t, i.e., $l \leq k$. Following Lemma 2.4 and after some straightforward algebraic operations, we have

$$\begin{aligned} &W^{\mathcal{A}}_t(\omega_1, \cdots, \omega'_l, \cdots, \omega_N) - W^{\mathcal{A}}_t(\omega_1, \cdots, \omega_l, \cdots, \omega_N) \\ &\quad = (\omega'_l - \omega_l)[W^{\mathcal{A}}_t(\omega_1, \cdots, 1, \cdots, \omega_N) - W^{\mathcal{A}}_t(\omega_1, \cdots, 0, \cdots, \omega_N)] \end{aligned}$$

Let $\mathcal{M} \triangleq \mathcal{N}(k) \backslash \{l\} = \{1, \cdots, l-1, l+1, \cdots, k\}$, by developing $W^{\mathcal{A}}_t(\Omega(t))$ as a function of ω_l, we have

$$
\begin{aligned}
W_t^{\mathcal{A}}(\Omega(t)) = &\; U(\omega_1(t), \cdots, \omega_k(t)) \\
&+ \beta(1-\mathcal{E})\omega_l \sum_{\mathcal{E} \subseteq \mathcal{M}} \Pr(\mathcal{M}, \mathcal{E}) \widehat{W_{t+1}^{\mathcal{A}}}(\Omega_{\mathcal{E}}(t+1)) \\
&+ \beta\Big[1-(1-\mathcal{E})\omega_l\Big] \sum_{\mathcal{E} \subseteq \mathcal{M}} \Pr(\mathcal{M}, \mathcal{E}) \widehat{W_{t+1}^{\mathcal{A}}}(\Omega'_{\mathcal{E}}(t+1))
\end{aligned}
$$

Let $\omega_l = 0$ and 1, respectively, we have

$$
W_t^{\mathcal{A}}(\omega_1, \cdots, 0, \cdots, \omega_N) = U(\omega_1, \cdots, 0, \cdots, \omega_N) + \beta \sum_{\mathcal{E} \subseteq \mathcal{M}} \Pr(\mathcal{M}, \mathcal{E}) \widehat{W_{t+1}^{\mathcal{A}}}(\Omega_0^{\mathcal{E}}(t+1)),
$$

$$
\begin{aligned}
W_t^{\mathcal{A}}(\omega_1, \cdots, 1, \cdots, \omega_N) = &\; U(\omega_1, \cdots, 1, \cdots, \omega_N) \\
&+ \beta(1-\mathcal{E}) \sum_{\mathcal{E} \subseteq \mathcal{M}} \Pr(\mathcal{M}, \mathcal{E}) \widehat{W_{t+1}^{\mathcal{A}}}(\Omega_{1-\mathcal{E}}^{\mathcal{E}}(t+1)) \\
&+ \beta\mathcal{E} \sum_{\mathcal{E} \subseteq \mathcal{M}} \Pr(\mathcal{M}, \mathcal{E}) \widehat{W_{t+1}^{\mathcal{A}}}(\Omega_{\mathcal{E}}^{\mathcal{E}}(t+1))
\end{aligned}
$$

where

$$
\begin{aligned}
\Omega_0^{\mathcal{E}}(t+1) &= (\mathbf{P}_{11}^{\mathcal{E}}, \Phi(k+1, N), \bar{\mathbf{Q}}^{\mathcal{M},\mathcal{E},l}, p_{01}, \underline{\mathbf{Q}}^{\mathcal{M},\mathcal{E},l}) \\
\Omega_{1-\mathcal{E}}^{\mathcal{E}}(t+1) &= (\mathbf{P}_{11}^{\mathcal{E}}, p_{11}, \Phi(k+1, N), \bar{\mathbf{Q}}^{\mathcal{M},\mathcal{E},l}, \underline{\mathbf{Q}}^{\mathcal{M},\mathcal{E},l}) \\
\Omega_{\mathcal{E}}^{\mathcal{E}}(t+1) &= \left(\mathbf{P}_{11}^{\mathcal{E}}, \Phi(k+1, N), \overline{\mathbf{Q}}^{\mathcal{M},\mathcal{E},l}, p_{11}, \underline{\mathbf{Q}}^{\mathcal{M},\mathcal{E},l}\right)
\end{aligned}
$$

It can be checked that $\Omega_{1-\mathcal{E}}^{\mathcal{E}}(t+1) \geq \Omega_0^{\mathcal{E}}(t+1)$ and $\Omega_{\mathcal{E}}^{\mathcal{E}}(t+1) \geq \Omega_0^{\mathcal{E}}(t+1)$. It then follows from induction that given $\mathcal{E}$, $\widehat{W_{t+1}^{\mathcal{A}}}(\Omega_{1-\mathcal{E}}^{\mathcal{E}}(t+1)) \geq \widehat{W_{t+1}^{\mathcal{A}}}(\Omega_0^{\mathcal{E}}(t+1))$ and $\widehat{W_{t+1}^{\mathcal{A}}}(\Omega_{\mathcal{E}}^{\mathcal{E}}(t+1)) \geq \widehat{W_{t+1}^{\mathcal{A}}}(\Omega_0^{\mathcal{E}}(t+1))$. Noticing that $U(\bullet)$ is increasing in each element, we then have

$$
\begin{aligned}
& W_t^{\mathcal{A}}(\omega_1, \cdots, 1, \cdots, \omega_N) - W_t^{\mathcal{A}}(\omega_1, \cdots, 0, \cdots, \omega_N) \\
&\quad = U(\omega_1, \cdots, 1, \cdots, \omega_N) - U(\omega_1, \cdots, 0, \cdots, \omega_N) \\
&\quad + \beta(1-\mathcal{E}) \sum_{\mathcal{E} \subseteq \mathcal{M}} \Pr(\mathcal{M}, \mathcal{E}) \Big[\widehat{W_{t+1}^{\mathcal{A}}}(\Omega_{1-\mathcal{E}}^{\mathcal{E}}(t+1)) - \widehat{W_{t+1}^{\mathcal{A}}}(\Omega_0^{\mathcal{E}}(t+1))\Big] \\
&\quad + \beta\mathcal{E} \sum_{\mathcal{E} \subseteq \mathcal{M}} \Pr(\mathcal{M}, \mathcal{E}) \Big[\widehat{W_{t+1}^{\mathcal{A}}}(\Omega_{\mathcal{E}}^{\mathcal{E}}(t+1)) - \widehat{W_{t+1}^{\mathcal{A}}}(\Omega_0^{\mathcal{E}}(t+1))\Big] \\
&\quad \geq 0
\end{aligned}
$$

Combining the above analysis in two cases completes our proof.

Proof of Lemmas 2.6 and 2.7

Due to the dependency between the two lemmas, we prove them together by backward induction.

We first show that Lemmas 2.6 and 2.7 hold for slot T. It is easy to verify that Lemma 2.6 holds. We then prove Lemma 2.7. Noticing that $p_{01} \leq \omega_N \leq \omega_k \leq p_{11} \leq 1$, we have

$$\begin{aligned}
&W_T^{\mathcal{A}}(\omega_1, \cdots, \omega_N) - W_T^{\mathcal{A}}(\omega_N, \omega_1, \cdots, \omega_{N-1}) \\
&= U(\omega_1, \cdots, \omega_k) - U(\omega_N, \omega_1, \cdots, \omega_{k-1}) \\
&= (\omega_k - \omega_N)[U(\omega_1, \cdots, \omega_{k-1}, 1) - U(\omega_1, \cdots, \omega_{k-1}, 0)] \leq (1 - \omega_N)\Delta_{\max}
\end{aligned}$$

$$\begin{aligned}
&W_T^{\mathcal{A}}(\omega_1, \cdots, \omega_N) - W_T^{\mathcal{A}}(\omega_N, \omega_2, \cdots, \omega_{N-1}, \omega_1) \\
&= U(\omega_1, \cdots, \omega_k) - U(\omega_N, \omega_2, \cdots, \omega_{k-1}) \\
&= (\omega_1 - \omega_N)[U(1, \omega_2, \cdots, \omega_k) - U(0, \omega_2, \cdots, \omega_k)] \leq (p_{11} - p_{01})\Delta_{\max}
\end{aligned}$$

Lemma 2.7 thus holds for slot T.

Assume that Lemmas 2.6 and 2.7 hold for slots $T, \cdots, t+1$, we now prove that it holds for slot t.

We first prove Lemma 2.6. Considering $l < m$, we distinguish the following three cases.

Case 1: $l \geq k+1, \mathcal{A}(t) = \{1, 2, \cdots, k\}$. In this case, we have

$$\begin{aligned}
&W_l^{\mathcal{A}}(\omega_1, \cdots, \omega_l, \cdots, \omega_m, \cdots, \omega_N) - W_l^{\mathcal{A}}(\omega_1, \cdots, \omega_m, \cdots, \omega_l, \cdots, \omega_N) \\
&= (\omega_l - \omega_m)[W_t^{\mathcal{A}}(\omega_1, \cdots, 1, \cdots, 0, \cdots, \omega_N) - W_t^{\mathcal{A}}(\omega_1, \cdots, 0, \cdots, 1, \cdots, \omega_N)] \\
&= (\omega_l - \omega_m)\beta \sum_{\mathcal{E} \subseteq \mathcal{A}(t)} \Pr(\mathcal{A}(t), \mathcal{E})[\widehat{W_{t+1}^{\mathcal{A}}}(\Omega_{\mathcal{E}}(t+1)) - \widehat{W_{t+1}^{\mathcal{A}}}(\Omega'_{\mathcal{E}}(t+1))]
\end{aligned}$$

where

$$\Omega_{\mathcal{E}}(t+1) = \left(\mathbf{P}_{11}^{\mathcal{E}}, \Gamma(\omega_{k+1}), \cdots, p_{11}, \cdots, p_{01}, \cdots, \Gamma(\omega_N), \mathbf{Q}^{\mathcal{A}(t),\mathcal{E}}\right)$$

$$\Omega'_{\mathcal{E}}(t+1) = \left(\mathbf{P}_{11}^{\mathcal{E}}, \Gamma(\omega_{k+1}), \cdots, p_{01}, \cdots, p_{11}, \cdots, \Gamma(\omega_N), \mathbf{Q}^{\mathcal{A}(t),\mathcal{E}}\right)$$

It follows from the induction result that $\widehat{W_{t+1}^{\mathcal{A}}}(\Omega_{\mathcal{E}}(t+1)) \geq \widehat{W_{t+1}^{\mathcal{A}}}\left(\Omega'_{\mathcal{E}}(t+1)\right)$. Hence

$$W_t^{\mathcal{A}}(\omega_1, \cdots, \omega_l, \cdots, \omega_m, \cdots, \omega_N) \geq W_t^{\mathcal{A}}(\omega_1, \cdots, \omega_m, \cdots, \omega_l, \cdots, \omega_N)$$

Case 2: $l \leq k$ and $m \geq k+1$ In this case, denote $\mathcal{M} \triangleq \mathcal{N}(k) \backslash \{l\}$, it can be noted that $\mathbf{Q}^{\mathcal{M},\mathcal{E}} = \mathbf{Q}^{\mathcal{M},\mathcal{E},l} + \bar{\mathbf{Q}}^{\mathcal{M},\mathcal{E},l}$. In this case, we have

$$W_t^{\mathcal{A}}(\omega_1,\cdots,\omega_l,\cdots,\omega_m,\cdots,\omega_N) - W_t^{\mathcal{A}}(\omega_1,\cdots,\omega_m,\cdots,\omega_l,\cdots,\omega_N)$$
$$= (\omega_l - \omega_m)\big[W_t^{\mathcal{A}}(\omega_1,\cdots,1,\cdots,0,\cdots,\omega_N) - W_t^{\mathcal{A}}(\omega_1,\cdots,0,\cdots,1,\cdots,\omega_N)\big]$$
$$= (\omega_l - \omega_m)[U(\omega_1,\cdots,1,\cdots,\omega_k) - U(\omega_1,\cdots,0,\cdots,\omega_k)$$
$$+\beta\sum_{\mathcal{E}\subseteq\mathcal{M}} \Pr(\mathcal{M},\mathcal{E})[(1-\mathcal{E})\widehat{W_{t+1}^{\mathcal{A}}}(\mathbf{P}_{11}^{\mathcal{E}},p_{11},\Gamma(\omega_{k+1}),\cdots,p_{01},\cdots,\Gamma(\omega_N),\mathbf{Q}^{\mathcal{M},\mathcal{E}})$$
$$+\mathcal{E}\widehat{W_{t+1}^{\mathcal{A}}}(\mathbf{P}_{11}^{\mathcal{E}},\Gamma(\omega_{k+1}),\cdots,p_{01},\cdots,\Gamma(\omega_N),\bar{\mathbf{Q}}^{\mathcal{M},\mathcal{E},l},p_{11},\underline{\mathbf{Q}}^{\mathcal{M},\mathcal{E},l})$$
$$-\widehat{W_{t+1}^{\mathcal{A}}}(\mathbf{P}_{11}^{\mathcal{E}},\Gamma(\omega_{k+1}),\cdots,p_{11},\cdots,\Gamma(\omega_N),\bar{\mathbf{Q}}^{\mathcal{M},\mathcal{E},l},p_{01},\underline{\mathbf{Q}}^{\mathcal{M},\mathcal{E},l})]]$$
$$\geq (\omega_l - \omega_m)[\Delta_{\min} + \beta\sum_{\mathcal{E}\subseteq\mathcal{M}} \Pr(\mathcal{M},\mathcal{E})$$
$$\cdot[(1-\mathcal{E})\widehat{W_{t+1}^{\mathcal{A}}}(p_{01},\mathbf{P}_{11}^{\mathcal{E}},p_{11},\Gamma(\omega_{k+1}),\cdots,\Gamma(\omega_N),\mathbf{Q}^{\mathcal{M},\mathcal{E}})$$
$$+\mathcal{E}\widehat{W_{t+1}^{\mathcal{A}}}(p_{01},\mathbf{P}_{11}^{\mathcal{E}},\Gamma(\omega_{k+1}),\cdots,\Gamma(\omega_N),\mathbf{Q}^{\mathcal{M},\mathcal{E}},p_{11})$$
$$-\widehat{W_{t+1}^{\mathcal{A}}}(\mathbf{P}_{11}^{\mathcal{E}},p_{11},\Gamma(\omega_{k+1}),\cdots,\Gamma(\omega_N),\mathbf{Q}^{\mathcal{M},\mathcal{E}},p_{01})]]$$
$$\geq (\omega_l - \omega_m)\left[\Delta_{\min} - \beta\sum_{\mathcal{E}\subseteq\mathcal{M}} \Pr(\mathcal{M},\mathcal{E})\right.$$
$$\left.\times\times\left((1-\mathcal{E})(1-p_{01})\Delta_{\max} + \mathcal{E}(p_{11}-p_{01})\Delta_{\max}\frac{1-[\beta(1-\mathcal{E})(p_{11}-p_{01})]^{T-t}}{1-\beta(1-\mathcal{E})(p_{11}-p_{01})}\right)\right]$$
$$\geq (\omega_l - \omega_m)\sum_{\mathcal{E}\subseteq\mathcal{M}} \Pr(\mathcal{M},\mathcal{E})$$
$$\times\left[\Delta_{\min} - \beta\left((1-\mathcal{E})(1-p_{01})\Delta_{\max} + \mathcal{E}(p_{11}-p_{01})\Delta_{\max}\frac{1}{1-(1-\mathcal{E})(p_{11}-p_{01})}\right)\right]$$
$$\geq 0$$

where the first inequality follows the induction result of Lemma 2.6, the second inequality follows the induction result of Lemma 2.7, the third inequality follows the condition in the lemma.

Case 3: l, $m \geq k$. This case follows Lemma 2.3.

Combing the above three cases, thus lemma 2.6 is proven for slot t.

We then proceed to prove Lemma 2.7. We start with the first inequality. We develop W_t w.r.t. ω_k and ω_N according to Lemma 2.4 as follows:

$$W_t^{\mathcal{A}}(\omega_1,\cdots,\omega_{k-1},\omega_k,\cdots,\omega_{n-1},\omega_N) - W_t^{\mathcal{A}}(\omega_N,\omega_1,\cdots,\omega_{k-1},\omega_k,\cdots,\omega_{N-1})$$
$$= \omega_k\omega_N \times \underbrace{[W_t^{\mathcal{A}}(\omega_1,\cdots,\omega_{k-1},1,\omega_{k+1},\cdots,\omega_{N-1},1) - W_t^{\mathcal{A}}(1,\omega_1,\cdots,\omega_{k-1},1,\omega_{k+1},\cdots,\omega_{N-1})]}_{termA}$$
$$+\omega_k(1-\omega_N) \times \underbrace{[W_t^{\mathcal{A}}(\omega_1,\cdots,\omega_{k-1},1,\omega_{k+1},\cdots,\omega_{N-1},0) - W_t^{\mathcal{A}}(0,\omega_1,\cdots,\omega_{k-1},1,\omega_{k+1},\cdots,\omega_{N-1})]}_{termB}$$
$$+(1-\omega_k)\omega_N \times \underbrace{[W_t^{\mathcal{A}}(\omega_1,\cdots,\omega_{k-1},0,\omega_{k+1},\cdots,\omega_{N-1},1) - W_t^{\mathcal{A}}(1,\omega_1,\cdots,\omega_{k-1},0,\omega_{k+1},\cdots,\omega_{N-1})]}_{termC}$$
$$+(1-\omega_k)(1-\omega_N) \times \underbrace{[W_t^{\mathcal{A}}(\omega_1,\cdots,\omega_{k-1},0,\omega_{k+1},\cdots,\omega_{N-1},0) - W_t^{\mathcal{A}}(0,\omega_1,\cdots,\omega_{k-1},0,\omega_{k+1},\cdots,\omega_{n-1})]}_{termD}$$

We proceed the proof by upbounding the four terms in (2.39).

For the first term A, we have

$$\begin{aligned}
&W_t^{\mathcal{A}}(\omega_1,\cdots,\omega_{k-1},1,\omega_{k+1},\cdots,\omega_{N-1},1) - W_t^{\mathcal{A}}(1,\omega_1,\cdots,\omega_{k-1},1,\omega_{k+1},\cdots,\omega_{N-1}) \\
&= \beta \sum_{\mathcal{E} \subseteq \mathcal{N}(k-1)} \Pr(\mathcal{N}(k-1),\mathcal{E}) \\
&\times \Big[(1-\mathcal{E})\widehat{W_{t+1}^{\mathcal{A}}}\Big(\mathbf{P}_{11}^{\mathcal{E}}, p_{11}, \Phi(k+1,N-1), p_{11}, \mathbf{Q}^{\mathcal{N}(k-1),\mathcal{E}}\Big) \\
&+\mathcal{E}\widehat{W_{t+1}^{A}}\Big(\mathbf{P}_{11}^{\mathcal{E}}, \Phi(k+1,N-1), p_{11}, \mathbf{Q}^{\mathcal{N}(k-1),\mathcal{E}}, p_{11}\Big) \\
&-(1-\mathcal{E})\widehat{W_{t+1}^{\mathcal{A}}}\Big(p_{11}, \mathbf{P}_{11}^{\mathcal{E}}, p_{11}, \Phi(k+1,N-1), \mathbf{Q}^{\mathcal{N}(k-1),\mathcal{E}}\Big) \\
&-\mathcal{E}\widehat{W_{t+1}^{\mathcal{A}}}\Big(\mathbf{P}_{11}^{\mathcal{E}}, p_{11}, \Phi(k+1,N-1), p_{11}, \mathbf{Q}^{\mathcal{N}(k-1),\mathcal{E}}\Big)\Big] \le 0
\end{aligned}$$

where, the inequality follows the induction of Lemma 2.6.

For the second term B, we have

$$\begin{aligned}
&W_t^{\mathcal{A}}(\omega_1,\cdots,\omega_{k-1},1,\omega_{k+1},\cdots,\omega_{n-1},0) - W_t^{\mathcal{A}}(0,\omega_1,\cdots,\omega_{k-1},1,\omega_{k+1},\cdots,\omega_{n-1}) \\
&= U(\omega_1,\cdots,\omega_{k-1},1) - U(0,\omega_1,\cdots,\omega_{k-1}) + \beta \sum_{\mathcal{E} \subseteq \mathcal{N}(k-1)} \Pr(\mathcal{N}(k-1),\mathcal{E}) \\
&\times \Big[(1-\mathcal{E})\widehat{W_{t+1}^{\mathcal{A}}}\Big(\mathbf{P}_{11}^{\mathcal{E}}, p_{11}, \Phi(k+1,N-1), p_{01}, \mathbf{Q}^{\mathcal{N}(k-1),\mathcal{E}}\Big) \\
&+\mathcal{E}\widehat{W_{t+1}^{\mathcal{A}}}\Big(\mathbf{P}_{11}^{\mathcal{E}}, \Phi(k+1,N-1), p_{01}, \mathbf{Q}^{\mathcal{N}(k-1),\mathcal{E}}, p_{11}\Big) \\
&-\widehat{W_{t+1}^{\mathcal{A}}}\Big(\mathbf{P}_{11}^{\mathcal{E}}, p_{11}, \Phi(k+1,N-1), p_{01}, \mathbf{Q}^{\mathcal{N}(k-1),\mathcal{E}}\Big)\Big]
\end{aligned}$$

$$
\begin{aligned}
&= U(\omega_1,\cdots,\omega_{k-1},1) - U(0,\omega_1,\cdots,\omega_{k-1}) + \beta \sum_{\mathcal{E}\subseteq \mathcal{N}(k-1)} \Pr(\mathcal{N}(k-1),\mathcal{E}) \\
&\times\Big[\mathcal{E}\widehat{W_{t+1}^{\mathcal{A}}}\Big(\mathbf{P}_{11}^{\mathcal{E}},\Phi(k+1,N-1),p_{01},\mathbf{Q}^{\mathcal{N}(k-1),\mathcal{E}},p_{11}\Big) \\
&-\mathcal{E}\widehat{W_{t+1}^{\mathcal{A}}}\Big(\mathbf{P}_{11}^{\mathcal{E}},p_{11},\Phi(k+1,N-1),p_{01},\mathbf{Q}^{\mathcal{N}(k-1),\mathcal{E}}\Big)\Big] \\
&\leq \Delta_{\max}
\end{aligned}
$$

following the induction of Lemma 2.6.

For the third term C, we have

$$
\begin{aligned}
&W_t^{\mathcal{A}}(\omega_1,\cdots,\omega_{k-1},0,\omega_{k+1},\cdots,\omega_{N-1},1) - W_t^{\mathcal{A}}(1,\omega_1,\cdots,\omega_{k-1},0,\omega_{k+1},\cdots,\omega_{N-1}) \\
&= U(\omega_1,\cdots,\omega_{k-1},0) - U(1,\omega_1,\cdots,\omega_{k-1}) + \beta \sum_{\mathcal{E}\subseteq \mathcal{N}(k-1)} \Pr(\mathcal{N}(k-1),\mathcal{E}) \\
&\times\Big[\widehat{W_{t+1}^{\mathcal{A}}}\Big(\mathbf{P}_{11}^{\mathcal{E}},\Phi(k+1,N-1),p_{11},\mathbf{Q}^{\mathcal{N}(k-1),\mathcal{E}},p_{01}\Big) \\
&-(1-\mathcal{E})\widehat{W_{t+1}^{\mathcal{A}}}\Big(p_{11},\mathbf{P}_{11}^{\mathcal{E}},p_{01},\Phi(k+1,N-1),\mathbf{Q}^{\mathcal{N}(k-1),\mathcal{E}}\Big) \\
&-\mathcal{E}\widehat{W_{t+1}^{\mathcal{A}}}\Big(\mathbf{P}_{11}^{\mathcal{E}},p_{01},\Phi(k+1,N-1),p_{11},\mathbf{Q}^{\mathcal{N}(k-1),\mathcal{E}}\Big)\Big] \\
&\leq -\Delta_{\min} \\
&+\beta\sum_{\mathcal{E}\subseteq \mathcal{N}(k-1)} \Pr(\mathcal{N}(k-1),\mathcal{E})\Big[\widehat{W_{t+1}^{\mathcal{A}}}(\mathbf{P}_{11}^{\mathcal{E}},p_{11},\Phi(k+1,N-1),\mathbf{Q}^{\mathcal{N}(k-1),\mathcal{E}},p_{01}) \\
&-(1-\mathcal{E})\widehat{W_{t+1}^{\mathcal{A}}}(p_{01},p_{11},\mathbf{P}_{11}^{\mathcal{E}},\Phi(k+1,N-1),\mathbf{Q}^{\mathcal{N}(k-1),\mathcal{E}}) \\
&-\mathcal{E}\widehat{W_{t+1}^{\mathcal{A}}}(p_{01},\mathbf{P}_{11}^{\mathcal{E}},\Phi(k+1,N-1),\mathbf{Q}^{\mathcal{N}(k-1),\mathcal{E}},p_{11}) \\
&\leq -\Delta_{\min} + \beta\sum_{\mathcal{E}\subseteq \mathcal{N}(k-1)} \Pr(\mathcal{N}(k-1),\mathcal{E}) \\
&\times\left[(1-\mathcal{E})(1-p_{01})\Delta_{\max} + \mathcal{E}(p_{11}-p_{01})\Delta_{\max}\frac{1-[\beta(1-\mathcal{E})(p_{11}-p_{01})]^{T-t}}{1-\beta(1-\mathcal{E})(p_{11}-p_{01})}\right] \\
&\leq \sum_{\mathcal{E}\subseteq \mathcal{N}(k-1)} \Pr(\mathcal{N}(k-1),\mathcal{E}) \\
&\times\left[-\Delta_{\min} + \beta\left[(1-\mathcal{E})(1-p_{01})\Delta_{\max} + \frac{\mathcal{E}(p_{11}-p_{01})\Delta_{\max}}{1-(1-\mathcal{E})(p_{11}-p_{01})}\right]\right] \\
&\leq 0
\end{aligned}
$$

where the first inequality follows the induction result of Lemma 2.6, the second equality follows the induction result of Lemma 2.7, the fourth inequality is due the condition in Lemma 2.7.

For the fourth term D, we have

$$\begin{aligned}
&W_t^{\mathcal{A}}(\omega_1,\cdots,\omega_{k-1},0,\omega_{k+1},\cdots,\omega_{N-1},0)-W_t^{\mathcal{A}}(0,\omega_1,\cdots,\omega_{k-1},0,\omega_{k+1},\cdots,\omega_{N-1})\\
&=\beta\sum_{\mathcal{E}\subseteq\mathcal{N}(k-1)}\Pr(\mathcal{N}(k-1),\mathcal{E})\Big[W_{t+1}^{\widehat{\mathcal{A}}}\Big(\mathbf{P}_{11}^{\mathcal{E}},\Phi(k+1,N-1),p_{01},\mathbf{Q}^{\mathcal{N}(k-1),\mathcal{E}},p_{01}\Big)\\
&\qquad-W_{t+1}^{\widehat{\mathcal{A}}}\Big(\mathbf{P}_{11}^{\mathcal{E}},p_{01},\Phi(k+1,N-1),\mathbf{Q}^{\mathcal{N}(k-1),\mathcal{E}},p_{01}\Big)\Big]\\
&=\beta\sum_{\mathcal{E}\subseteq\mathcal{N}(k-1)}\Pr(\mathcal{N}(k-1),\mathcal{E})\Big[W_{t+1}^{\widehat{\mathcal{A}}}\Big(\mathbf{P}_{11}^{\mathcal{E}},\Phi(k+1,N-1),p_{01},\mathbf{Q}^{\mathcal{N}(k-1),\mathcal{E}},p_{01}\Big)\\
&\qquad-W_{t+1}^{\widehat{\mathcal{A}}}\Big(p_{01},\mathbf{P}_{11}^{\mathcal{E}},\Phi(k+1,N-1),\mathbf{Q}^{\mathcal{N}(k-1),\mathcal{E}},p_{01}\Big)\Big]\\
&\leqslant\beta\sum_{\mathcal{E}\subseteq\mathcal{N}(k-1)}\Pr(\mathcal{N}(k-1),\mathcal{E})\Big[W_{t+1}^{\widehat{\mathcal{A}}}\Big(\mathbf{P}_{11}^{\mathcal{E}},\Phi(k+1,N-1),\mathbf{Q}^{\mathcal{N}(k-1),\mathcal{E}},p_{01},p_{01}\Big)\\
&\qquad-W_{t+1}^{\widehat{\mathcal{A}}}\Big(p_{01},\mathbf{P}_{11}^{\mathcal{E}},\Phi(k+1,N-1),\mathbf{Q}^{\mathcal{N}(k-1),\mathcal{E}},p_{01}\Big)\Big]\le(1-p_{01})\beta\Delta_{\max}
\end{aligned}$$

where, the second equality follows Lemma 2.3, the first inequality follows the induction result of Lemma 2.6 and the second inequality follows the induction result of Lemma 2.7.

Combing the above results of the four terms, we have

$$\begin{aligned}
&W_t^{\mathcal{A}}(\omega_1,\cdots,\omega_N)-W_t^{\mathcal{A}}(\omega_N,\omega_1,\cdots,\omega_{N-1})\\
&\le\omega_k(1-\omega_N)\cdot\Delta_{\max}+(1-\omega_k)(1-\omega_N)\cdot(1-p_{01})\beta\Delta_{\max}\\
&\le\omega_k(1-\omega_N)\Delta_{\max}+(1-\omega_k)(1-\omega_N)\Delta_{\max}\le(1-\omega_N)\Delta_{\max}
\end{aligned}$$

which completes the proof of the first part of Lemma 2.7.

Finally, we prove the second part of Lemma 2.7. To this end, denote $\mathcal{M}\triangleq\{2,\cdots,k\}$, we have

$$\begin{aligned}
&W_t(\omega_1,\omega_2,\cdots,\omega_{N-1},\omega_N)-W_t(\omega_N,\omega_2,\cdots,\omega_{N-1},\omega_1)\\
&=(\omega_1-\omega_N)[W_t(1,\omega_2,\cdots,\omega_{N-1},0)-W_t(0,\omega_2,\cdots,\omega_{N-1},1)]\\
&=(\omega_1-\omega_N)[F(1,\omega_2,\cdots,\omega_k)-F(0,\omega_2,\cdots,\omega_k)\\
&+\beta\sum_{\mathcal{E}\subseteq\mathcal{M}}\Pr(\mathcal{M},\mathcal{E})[(1-\mathcal{E})W_{t+1}(\mathbf{P}_{11}^{\mathcal{E}},p_{11},\Phi(k+1,N-1),p_{01},\mathbf{Q}^{\mathcal{M},\mathcal{E}})
\end{aligned}$$

$$
\begin{aligned}
&+\mathcal{E}W_{t+1}(\mathbf{P}_{11}^{\mathcal{E}}, \Phi(k+1,N-1), p_{01}, p_{11}, \mathbf{Q}^{\mathcal{M},\mathcal{E}}) \\
&-W_{t+1}(\mathbf{P}_{11}^{\mathcal{E}}, \Phi(k+1,N-1), p_{11}, p_{01}, \mathbf{Q}^{\mathcal{M},\mathcal{E}})] \\
&\leq (\omega_1 - \omega_N)[\Delta_{\max} \\
&+\beta \sum_{\mathcal{E} \subseteq \mathcal{M}} \Pr(\mathcal{M}, \mathcal{E}) \big[(1-\mathcal{E})W_{t+1}\big(\mathbf{P}_{11}^{\mathcal{E}}, p_{11}, \Phi(k+1,N-1), p_{01}, \mathbf{Q}^{\mathcal{M},\mathcal{E}}\big) \\
&+\mathcal{E}W_{t+1}(\mathbf{P}_{11}^{\mathcal{E}}, \Phi(k+1,N-1), p_{01}, p_{11}, \mathbf{Q}^{\mathcal{M},\mathcal{E}}) \\
&-W_{t+1}\big(\mathbf{P}_{11}^{\mathcal{E}}, \Phi(k+1,N-1), p_{01}, p_{11}, \mathbf{Q}^{\mathcal{M},\mathcal{E}}\big)\big] \\
&= (\omega_1 - \omega_N)[\Delta_{\max} \\
&+\beta \sum_{\mathcal{E} \subseteq \mathcal{M}} \Pr(\mathcal{M}, \mathcal{E})[(1-\mathcal{E})W_{t+1}(\mathbf{P}_{11}^{\mathcal{E}}, p_{11}, \Phi(k+1,N-1), p_{01}, \mathbf{Q}^{\mathcal{M},\mathcal{E}}) \\
&-(1-\mathcal{E})W_{t+1}(\mathbf{P}_{11}^{\mathcal{E}}, \Phi(k+1,N-1), p_{01}, p_{11}, \mathbf{Q}^{\mathcal{M},\mathcal{E}})]] \\
&\leq (\omega_1 - \omega_N)[\Delta_{\max} \\
&+\beta \sum_{\mathcal{E} \subseteq \mathcal{M}} \Pr(\mathcal{M}, \mathcal{E})[(1-\mathcal{E})W_{t+1}(\mathbf{P}_{\mathbf{11}}^{\mathcal{E}}, p_{11}, \mathbf{\Phi}(k+1,N-1), \mathbf{Q}^{\mathcal{M},\mathcal{E}}, p_{01}) \\
&-(1-\mathcal{E})W_{t+1}(p_{01}, \mathbf{P}_{11}^{\mathcal{E}}, \Phi(k+1,N-1), \mathbf{Q}^{\mathcal{M},\mathcal{E}}, p_{11})] \\
&\leq (p_{11} - p_{01})\Delta_{\max} \\
&+\beta \sum_{\mathcal{E} \subseteq \mathcal{M}} \Pr(\mathcal{M}, \mathcal{E})(1-\mathcal{E}) \frac{1 - [\beta(1-\mathcal{E})(p_{11}-p_{01})]^{T-t}}{1-\beta(1-\mathcal{E})(p_{11}-p_{01})}(p_{11}-p_{01})\Delta_{\max}] \\
&= \frac{1 - [\beta(1-\mathcal{E})(p_{11}-p_{01})]^{T-t+1}}{1-\beta(1-\mathcal{E})(p_{11}-p_{01})}(p_{11}-p_{01})\Delta_{\max}
\end{aligned}
$$

where the first two inequalities follow the recursive application of the induction result of Lemma 2.6, the third inequality follows the induction result of Lemma 2.7.

We thus complete the whole process of proving Lemmas 2.6 and 2.7.

References

1. P. Whittle, Restless bandits: Activity allocation in a changing world. J. Appl. Prob. **25A**, 287–298 (1988)
2. Q. Zhao, L. Tong, A. Swami, Y. Chen, Decentralized cognitive MAC for opportunistic spectrum access in ad hoc networks: A POMDP framework. IEEE J. Sel. Areas Commun. **25** (3), 589–600 (2007)
3. C.H. Papadimitriou, J.N. Tsitsiklis, The complexity of optimal queueing network control. Math. Oper. Res. **24**(2), 293–305 (1999)

4. R.R. Weber, G. Weiss, On an index policy for restless bandits. J. Appl. Prob. **27**(1), 637–648 (1990)
5. K. Liu, Q. Zhao, Indexability of restless bandit problems and optimality of whittle index for dynamic multichannel access. IEEE Trans. Inf. Theory **56**(11), 5547–5567 (2010)
6. Q. Zhao, B. Krishnamachari, K. Liu, On myopic sensing for multi-channel opportunistic access: Structure, optimality, and performance. IEEE Trans. Wirel. Commun. **7**(3), 54315440 (2008)
7. S. Ahmad, M. Liu, T. Javidi, Q. Zhao, B. Krishnamachari, Optimality of myopic sensing in multi-channel opportunistic access. IEEE Trans. Inf. Theory **55**(9), 4040–4050 (2009)
8. S. Ahmad, M. Liu. Multi-Channel Opportunistic Access: A Case of Restless Bandits with Multiple Plays. in *Allerton Conference*, Monticello, Il, Sep–Oct, 2009
9. K. Wang, L. Chen, On optimality of myopic policy for restless multi-armed bandit problem: An axiomatic approach. IEEE Trans. Signal Process. **60**(1), 300–309 (2012)
10. K. Wang, L. Chen, On the optimality of myopic sensing in multi-channel opportunistic access: The case of sensing multiple channels. IEEE Wireless Commun. Lett. **1**(5), 452–455 (2012)
11. Y. Chen, Q. Zhao, A. Swami, Joint design and separation principle for opportunistic spectrum access in the presence of sensing errors. IEEE Trans. Inf. Theory **54**(5), 2053–2071 (2008)
12. K. Liu, Q. Zhao, B. Krishnamachari, Dynamic multichannel access with imperfect channel state detection. IEEE Trans. Signal Process. **58**(5), 2795–2807 (2010)

Chapter 3
Whittle Index Policy for Opportunistic Scheduling: Heterogeneous Two-State Channels

3.1 Introduction

3.1.1 Background

We consider an opportunistic communication system with heterogenous.[1] Gilbert-Elliot channels [1], in which a user is limited to sense and transmit only on one channel each time due to limitation on sensing capability[2]. Given that channel sensing in practice is not perfect, the fundamental optimization problem addressed in this chapter is how the user exploits the imperfect sensing results and the stochastic properties of channels to maximize its utility (e.g., expected throughput) by switching among channels opportunistically.

3.1.2 Main Results and Contributions

The central pivot in the Whittle index policy analysis is to establish the indexability of the problem and compute the corresponding index. In our problem, for a subset of specific scenarios characterized by the corresponding parameter spaces (e.g., [2, 3]), the Whittle index policy degenerates to the myopic policy. However, beyond those scenarios, the structure of the index-based policy is still open, which is the focus of this paper (cf. Table 3.1).

[1]In Chap. 2, the homogeneous Gilbert-Elliot channels are studied.

[2]The technical analysis in this paper can be extended to address the case where a user is allowed to sense a fixed number of channels.

K. Wang, L. Chen, *Restless Multi-Armed Bandit in Opportunistic Scheduling*,
https://doi.org/10.1007/978-3-030-69959-8_3

Table 3.1 Summary of related work and this paper

Parameter domain	Policy	Optimality
$p_{11} \geqslant p_{01}, \varepsilon \leqslant \frac{p_{01}(1-p_{11})}{p_{11}(1-p_{01})}$	Myopic policy	Globally optimal [2]
$p_{11} \leqslant p_{01}, \varepsilon \leqslant \frac{p_{11}(1-p_{01})}{p_{01}(1-p_{11})}$	Myopic policy	Globally optimal [3]
$\varepsilon_i \leqslant \frac{\left(1-\max\left\{p_{11}^{(i)}, p_{01}^{(i)}\right\}\right)\cdot \min\left\{p_{11}^{(i)}, p_{01}^{(i)}\right\}}{\left(1-\min\left\{p_{11}^{(i)}, p_{01}^{(i)}\right\}\right)\cdot \max\left\{p_{11}^{(i)}, p_{01}^{(i)}\right\}}$	Index policy	Locally optimal [this chapter]

The major technical challenge to establish the indexability in our problem comes from the imperfect sensing, where the false alarm rate is involved in the propagation of belief information and makes the value function no longer linear as in existing studies. As a result, the traditional approach of computing the Whittle index cannot be used in this context. To the best of our knowledge, there does not exist a closed form Whittle index for the nonlinear case; only numerical simulation is conducted under a strict assumption on the indexability [4].

To address the challenge caused by nonlinearity, we investigate the fixed points of belief evolution function (which is nonlinear), based on which we establish a set of periodic structures of the resulting dynamic system. We then use the derived properties to linearize the value function by a piecewise approach to prove the Whittle indexability and derive the closed-form Whittle index. Our results in this paper thus solves the multichannel opportunistic scheduling problem under imperfect channel sensing by establishing its indexability and constructing the corresponding index policy. Due to the generality of the problem, our results can be applied in a wide range of engineering applications where the underlying optimization problems can be cast into restless bandits with imperfect sensing of bandit states. Therefore, the terminology and analysis in this paper should be understood generically.

3.2 Related Work

The opportunistic channel access can be cast into a RMAB problem, which is proved to be PSPACE-hard [5]. To the best of our knowledge, very few results are reported on the structure of the optimal policy of a generic RMAB due to its high complexity.

The myopic strategy, due to its simple and tractable structure, has recently attracted extensive research attention. It essentially consists of sensing the channels that maximize the expected immediate reward while ignoring the impact of the current decision on future reward. Along this research thrust, the optimality of the myopic policy is partially established for the homogeneous Gilbert-Elliot channel case under perfect sensing [6]. In [7], the authors studied the case of heterogeneous channels and derived a set of closed-form sufficient conditions to guarantee the optimality of the myopic policy. In [8], the authors proposed a sufficient condition framework for the optimality of myopic policy. In [9], the authors gave the sufficient

Table 3.2 Main notations

Symbols	Descriptions
$\mathcal{N}$	The set of N channels, i.e., 1, 2, . . ., N
$\boldsymbol{P}^{(i)}$	State transition matrix of channel i
T	The total number of time slots
t	Time slot index
ε_i	False alarm probability of channel i
ζ_i	Miss detection probability of channel i
$S_i(t)$	The state of channel i in slot t
$\omega_i(t)$	The conditional probability of being "good"
$\Omega(t)$	Channel state belief vector at slot t
$\Omega(0)$	The initial channel state belief vector
$O_i(t)$	The observation state of channel i
π_t	The mapping from $\Omega(t)$ to $A(t)$
$a_n(t)$	The action of channel n in slot t
β	Discount factor

conditions for multistate channels. For the imperfect sensing of Gilbert-Elliot channels, Liu et al. [10] proved the optimality of the myopic policy for the specifical case of two channels. In [2, 3, 11], the authors derived closed-form condition to guarantee the optimality of the myopic policy for arbitrary number of channels.

Generally speaking, the structure of the optimum access policy is only characterized for a subset of parameter space under which the myopic policy is proved optimum. Beyond this parameter space, we need to turn to a more generic policy, Whittle index policy, introduced by P. Whittle in [27]. The Whittle index policy has been a very popular heuristic for restless bandit, which, while suboptimal in general, is provably optimal in asymptotic sense [12, 13] and has good empirical performance. The Whittle index policy and its variants have been studied extensively in engineering applications, e.g., sensor scheduling [4, 14], multi-UAV coordination [15], crawling web content [16], channel allocation in wireless networks [17, 18], and job scheduling [19–21]. More comprehensive treatments of indexable restless bandits can be found in [22–24].

3.3 System Model

Table 3.2 summaries main notations used in this chapter.

We consider a time-slotted multichannel opportunistic communication system, in which a user is able to access a set $\mathcal{N}$ of N independent channels, each characterized by a Markov chain of two states, *good* (1) and *bad* (0). The channel state transition matrix $\mathbf{P}^{(i)}$ for channel $i (i \in \mathcal{N})$ is given as follows:

$$\mathbf{P}^{(i)} = \begin{bmatrix} 1 - p_{01}^{(i)} & p_{01}^{(i)} \\ 1 - p_{11}^{(i)} & p_{11}^{(i)} \end{bmatrix}$$

We assume that channels go through state transition at the beginning of each slot t. The system operates in a synchronously time-slotted fashion with the time slot indexed by $t(t = 0, 1, \cdots)$.

Due to hardware constraints and energy cost, the user is allowed to sense only one of the N channels at each slot t. We assume that the user makes the channel selection decision at the beginning of each slot after the channel state transition. Once a channel is selected, the user detects the channel state $S_i(t)$, which can be considered as a binary hypothesis test:

$$\mathcal{H}_0 : S_i(t) = 1 \text{ (good) vs.} \mathcal{H}_1 : S_i(t) = 0 \text{ (bad)}.$$

The performance of channel i state detection is characterized by the probability of false alarm ε_i and the probability of miss detection δ_i:

$$\varepsilon_i := \Pr\{ \text{ decide } \mathcal{H}_1 \mid \mathcal{H}_0 \text{ is true } \}$$
$$\delta_i := \Pr\{ \text{ decide } \mathcal{H}_0 \mid \mathcal{H}_1 \text{ is true } \}$$

Based on the imperfect detection outcome in slot t, the user determines whether t access channel i for transmission. We denote the action on channel n made by the user at slot t by $a_n(t)$, i.e.,

$$a_n(t) = \begin{cases} 1, & \text{if channel } n \text{ is chosen in slot } t \\ 0, & \text{if channel } n \text{ is not chosen in slot } t \end{cases}$$

Thus, $\sum_{n=1}^{N} a_n(t) = 1$ for all t, indicating that exactly one channel is chosen in each slot.

Since failed transmissions may occur, acknowledgments (ACKs) are necessary to ensure guaranteed delivery. Specifically, when the receiver successfully receives a packet from a channel, it sends an acknowledgment to the transmitter over the same channel at the end of slot. Otherwise, the receiver does nothing, i.e., a NAK is defined as the absence of an ACK, which occurs when the transmitter did not transmit over this channel or transmitted, but the channel is busy in this slot. We assume that acknowledgments are received without error since acknowledgments are always transmitted over idle channels.

Obviously, by imperfectly sensing only one of N channels, the user cannot observe the state information of the whole system. Hence, the user has to infer the channel states from its decision history and observation history so as to make its future decision. To this end, we define the *channel state belief vector* (hereinafter referred to as *belief vector* for briefness) $\Omega(t) := \{\omega_i(t), i \in \mathcal{N}\}$ where $0 \leq \omega_i(t) \leq 1$ is the conditional probability that channel i is in state good (i.e., $S_i(t) = 1$) conditioned on the decision history and observation history.

To ensure that the user and its intended receiver tune to the same channels in each slot, channel selections should be based on common observation: $K(t) \in \{0(\text{NAK}), 1(\text{ACK})\}$ in each slot rather than the detection outcome at the transmitter.

Given the sensing action $\{a_i(t)\}_{i\in\mathcal{N}}$ and the observation $K(t)$, the belief vector in $t + 1$ slot can be updated recursively using Bayes Rule as shown in (3.1):

$$\omega_i(t+1) = \begin{cases} p_{11}^{(i)}, & \text{if } a_i(t) = 1, K(t) = 1 \\ \Psi_i(\omega_i(t)), & \text{if } a_i(t) = 1, K(t) = 0 \\ \Gamma_i(\omega_i(t)), & \text{if } a_i(t) = 0 \end{cases} \tag{3.1}$$

where,

$$\Gamma_i(\omega_i(t)) := \omega_i(t)p_{11}^{(i)} + (1 - \omega_i(t))p_{01}^{(i)} \tag{3.2}$$

$$\varphi_i(\omega_i(t)) := \frac{\varepsilon_i \omega_i(t)}{1 - (1 - \varepsilon_i)\omega_i(t)}, \tag{3.3}$$

$$\Psi_i(\omega_i(t)) := \Gamma_i(\varphi_i(\omega_i(t))) \tag{3.4}$$

We would like to emphasize that the sensing error introduces technical complications in the system dynamics (i.e., $\varphi_i(\omega_i(t))$) due to its nonlinearity. Therefore, the analysis methods and results [7, 25, 26] in the perfect sensing case where the belief evolution is linear cannot be applied to the scenario with sensing error.

3.4 Problem Formulation

In this section, we formulate the optimization problem of opportunistic multichannel access faced by the user. Mathematically, let $\pi = \{\pi(t)\}_{t \geqslant 0}$ denote the sensing policy, with $\pi(t)$ defined as a mapping from the belief vector $\Omega(t)$ to the action of sensing one channel in each slot t:

$$\pi(t) : \Omega(t) \mapsto \{1, 2, \cdots, N\}, t = 0, 1, 2, \cdots \tag{3.5}$$

Let

$$a_n^{\pi}(t) = \begin{cases} 1, & \text{if channel } n \text{ is chosen under } \pi(t) \\ 0, & \text{if channel } n \text{ is not chosen under } \pi(t). \end{cases} \tag{3.6}$$

Let $\Pi_n := \{a_n^{\pi}(t) : t \geqslant 0\}$ be policy space on channel n under the sensing policy π, then $\Pi = \cup_{n=1}^{N} \Pi_n$ is the joint policy space.

We are interested in the user's optimization problem to find the optimal sensing policy π^* that maximizes the expected total discounted reward over an infinite horizon. The following gives the formal definition of the optimal sensing problem:

$$\text{OrigP}: \max_{\pi\in\Pi} \mathbb{E}\left\{\sum_{t=0}^{\infty}\beta^t \sum_{n=1}^{N}\left(a_n^\pi(t)\left(1-\varepsilon_n\right)\omega_n(t)\right)\right\} \tag{3.7}$$

$$\text{s.t.} \quad \sum_{n=1}^{N} a_n^\pi(t) = 1, t = 0, 1, \cdots, \infty \tag{3.8}$$

where the constraint (3.8) shows that only one channel can be chosen each time.

In the following, we decompose OrigP into N similar subproblems. By relaxing the constraint (3.8), we have

$$1 = \frac{\sum_{t=0}^{\infty}\beta^t \sum_{n=1}^{N} a_n^\pi(t)}{\sum_{t=0}^{\infty}\beta^t} = \frac{\sum_{n=1}^{N}\sum_{t=0}^{\infty}\beta^t a_n^\pi(t)}{\sum_{t=0}^{\infty}\beta^t} = \sum_{n=1}^{N}\frac{\sum_{t=0}^{\infty}\beta^t a_n^\pi(t)}{\sum_{t=0}^{\infty}\beta^t} \tag{3.9}$$

Divided by $\sum_{t=0}^{\infty}\beta^t$, we transform OrigP to the following relaxed problem RelxP.

$$RelxP: \max_{\pi\in\Pi} \mathbb{E}\sum_{n=1}^{N}\left\{\frac{\sum_{t=0}^{\infty}\beta^t (a_n^\pi(t)(1-\varepsilon_n)\omega_n(t))}{\sum_{t=0}^{\infty}\beta^t}\right\} \tag{3.10}$$

$$\text{s.t.} \quad \sum_{n=1}^{N}\frac{\sum_{t=0}^{\infty}\beta^t a_n^\pi(t)}{\sum_{t=0}^{\infty}\beta^t} = 1 \tag{3.11}$$

By introducing Lagrange multiplier ν, we can rewrite (3.10) as follows.

$$\max_{\pi\in\Pi} \mathbb{E}\sum_{n=1}^{N}\left\{\frac{\sum_{t=0}^{\infty}\beta^t (a_n^\pi(t)(1-\varepsilon_n)\omega_n(t)+\nu(1-a_n^\pi(t)))}{\sum_{t=0}^{\infty}\beta^t}\right\} \tag{3.12}$$

We further decompose (3.12) into N subproblems:

$$\max_{\pi_n \in \Pi_n} \mathbb{E}\left\{ \frac{\sum_{t=0}^{\infty} \beta^t (a_n^{\pi_n}(t)(1-\varepsilon_n)\omega_n(t) + \nu(1 - a_n^{\pi_n}(t)))}{\sum_{t=0}^{\infty} \beta^t} \right\} \tag{3.13}$$

Considering the constant $\beta(0 \leqslant \beta < 1)$, we have the following subproblem subP $- n$:

$$\text{subP} - n: \max_{\pi_n \in \Pi_n} \mathbb{E}\left\{ \sum_{t=0}^{\infty} \beta^t \left(a_n^{\pi_n}(t) \left(1 - \varepsilon_n\right) \omega_n(t) + \nu\left(1 - a_n^{\pi_n}(t)\right)\right) \right\} \tag{3.14}$$

To solve the original optimization problem OrigP, we first seek the optimal policy π_n^* for subproblem subP $- n(n \in \mathcal{N})$, and then construct a feasibly approximation policy $\pi = \left(\pi_1^*, \pi_2^*, \cdots, \pi_N^*\right)$ for the original problem OrigP.

3.5 Technical Preliminary: Indexability and Whittle Index

Let $V_{\beta,\ \nu}(\omega)$ be the value function corresponding to the subproblem (3.14), which denotes the maximum discounted reward accrued from a single-armed bandit process with subsidy v when the initial belief state is $\omega(0)$.

Considering two possible actions (choose or not) in each slot, we have

$$V_{\beta,v}(\omega) = \max\left\{V_{\beta,v}(\omega; a=0) V_{\beta,v}(\omega; a=1)\right\} \tag{3.15}$$

where,

$$V_{\beta,v}(\omega; a=0) = v + \beta V_{\beta,v}(\Gamma(\omega))$$

$$V_{\beta,v}(\omega; a=1) = (1-\varepsilon)\omega + \beta\Big[(1-\varepsilon)\omega V_{\beta,v}(p_{11}) + (1-(1-\varepsilon)\omega) V_{\beta,v}(\Psi(\omega))\Big]$$

$V_{\beta,\ \nu}(\omega; a = 1)$ denotes the reward obtained by taking action a in the first slot following by the optimal policy in future slots, and $V_{\beta,\ \nu}(\omega; a = 0)$ denotes the sum of the subsidy v obtained in the current slot under the passive action $(a = 0)$ and the total discounted future reward $\beta V_{\beta,\ \nu}(\Gamma(\omega))$.

Remark 3.1 In an infinite time horizon, a decision should be made at each slot, and the different decision leads to different evolution of belief information ω. Thus, in the following, we call (3.15) a dynamic system without introducing ambiguity.

Remark 3.2 We would like to point out that $V_{\beta,\ v}(\Psi(\omega))$ (specifically $\varphi(\omega)$) brings about the nonlinear belief update of the dynamic system (3.15), and leads to the complicated characteristics of the Whittle index.

The optimal action $a*$ for the belief state ω under subsidy v is given by

$$a^* = \begin{cases} 1, & \text{if } V_{\beta,v}(\omega; a=1) > V_{\beta,v}(\omega; a=0) \\ 0, & \text{otherwise.} \end{cases} \tag{3.16}$$

We define the passive set $\mathcal{P}(v)$ under subsidy v as

$$\mathcal{P}(v) := \left\{\omega : V_{\beta,v}(\omega; a=1) \leqslant V_{\beta,v}(\omega; a=0)\right\} \tag{3.17}$$

We next introduce some definitions related to the indexability of our problem.

Definition 3.1 (Indexability) Problem (3.14) is indexable if the passive set $\mathcal{P}(v)$ of the corresponding single-armed bandit process with subsidy v monotonically increases from ϕ to the whole state space [0, 1] as v increases from $-\infty$ to $+\infty$.

Under the indexability condition, Whittle index is defined as follows:

Definition 3.2 (Whittle index [27]) If Problem (3.14) is indexable, its Whittle index $W(\omega)$ of the state ω is the infimum subsidy v such that it is optimal to make the arm passive at ω. Equivalently, Whittle index is the infimum subsidy v that makes the passive and active actions equally rewarding

$$W(\omega) = \inf \left\{v : V_{\beta,v}(\omega; a=1) \leqslant V_{\beta,v}(\omega; a=0)\right\} \tag{3.18}$$

Definition 3.3 (Threshold Policy) Given a certain v, there exists $\omega^*(0 \leqslant \omega^* \leqslant 1)$ such that $V_{\beta,\ v}(\omega^*; 1) = V_{\beta,\ v}(\omega^*; 0)$. The threshold policy is defined as follows:

$$a^* = 1 \text{ for any } \omega\left(\omega^* < \omega \leqslant 1\right) \text{ while } a^* = 0 \text{ for any } \omega\left(0 \leqslant \omega < \omega^*\right),$$

or

$$a^* = 0 \text{ for any } \omega\left(\omega^* < \omega \leqslant 1\right) \text{ while } a^* = 1 \text{ for any } \omega\left(0 \leqslant \omega < \omega^*\right)$$

Definition 3.4 Problem (3.14) is CMI-indexable if the subsidy v computed by the threshold policy is a continuous and monotonically increasing (CMI) function of ω.

3.6 Whittle Index and Scheduling Rule

In this section, we summarize the main results of our paper. The detailed analysis and proofs of the results will be presented in later sections.

3.6.1 Whittle Index

Our central result is the establishment of the CMI indexability of the opportunistic multichannel access problem, as stated in the following theorem.

Theorem 3.1 *Given* $\varepsilon_i \leqslant \frac{(1-\max\{p_{11}^{(i)}, p_{01}^{(i)}\})\cdot \min\{p_{11}^{(i)}, p_{01}^{(i)}\}}{(1-\min\{p_{11}^{(i)}, p_{01}^{(i)}\})\cdot \max\{p_{11}^{(i)}, p_{01}^{(i)}\}} (\forall i \in \mathcal{N})$, *Problem* (3.14) *is CMI-indexable.*

To prove the indexability, we need to prove the continuity and increasing monotonicity of v in ω. Thus, we first derive the closed form v, and then prove that v is continuous and monotonically increasing in ω.

Based on the definition of threshold policy, we obtain the Whittle index in the following theorem.

Theorem 3.2 *The Whittle index* $W_\beta(\omega)$ *for channel i is given as follows.*

(i) *The case of positively correlated channels, i.e.,* $p_{11}^{(i)} \leqslant p_{01}^{(i)}$*: See* (3.19).
(ii) *The case of negatively correlated channels, i.e.,* $p_{11}^{(i)} \geqslant p_{01}^{(i)}$*: See* (3.20).

For the case of optimizing average reward, i.e., $\beta = 1$, we derive the Whittle index $W(\omega) = \lim_{\beta \to 1} W_\beta(\omega)$ as follows.

Theorem 3.3 *The Whittle index* $W(\omega)$ *for channel i is given as follows.*

(i) *The case of positively correlated channels, i.e.,* $p_{11}^{(i)} \leqslant p_{01}^{(i)}$ *: See* (3.21).
(ii) *The case of negatively correlated channels, i.e.,* $p_{11}^{(i)} \geqslant p_{01}^{(i)}$*: See* (3.22).

The following corollary bridges our results with existing body of works on myopic policy by showing that in a particular case with stochastically identical channels, the Whittle index-based policy we derive degenerates to the myopic policy.

Corollary 3.1 $W_\beta(\omega)$ *is a monotonically non-decreasing function of* ω*. As a consequence, the Whittle index policy is equivalent to the myopic (or greedy) policy for the considered RMAB with stochastically identical channels.*

3.6.2 Scheduling Rule

Based on the Whittle index, we can construct the index-based access policy for OrigP.

Each time to choose the channel $i^* = argmax_{i \in \mathcal{N}} W_\beta(\omega_i)$ for the discounted case and $i^* = argmax_{i \in \mathcal{N}} W(\omega_i)$ for the case of optimizing the average reward.

3.6.3 Technical Challenges

The main challenges in obtaining the indexability result in our problem comes from the nonlinear operator $\Psi_i(\cdot)$, summarized as follows:

1. The nonlinear operator $\Psi_i(\cdot)$ brings about nonlinear propagation of belief information in the evolution of the dynamic system.
2. The value function $V_{\beta,\nu}(\omega)$ is also nonlinear and intractable to compute due to the nonlinearity of $\Psi_i(\cdot)$.

To address the above challenges, we analyze the fixed points of the operators Γ_i, Ψ_i, as well as their combinations, and divide the belief information space into a series of regions using the fixed points. We then establish a set of periodic structures of the underlying nonlinear dynamic evolving system, based on which we further devise the linearization scheme for each region.

$$W_\beta(\omega)=\begin{cases}\dfrac{(1-\varepsilon_i)(\beta p_{01}^{(i)}+(\omega-\beta\Gamma_i(\omega)))}{1+\beta(p_{01}^{(i)}-\varepsilon_i p_{11}^{(i)})-\beta^2(1-\varepsilon_i)\Gamma_i(p_{11}^{(i)})-\beta(1-\varepsilon_i)(\omega-\Gamma_i(\omega))}, & \text{if}\, p_{11}^{(i)}\leqslant\omega<\omega_0^{(i)}\\ \dfrac{(1-\varepsilon_i)(\beta p_{01}^{(i)}+(1-\beta)\omega)}{1+\beta(p_{01}^{(i)}-\varepsilon_i p_{11}^{(i)})-\beta^2(1-\varepsilon_i)\Gamma_i(p_{11}^{(i)})-\beta(1-\beta)(1-\varepsilon_i)\omega}, & \text{if}\, \omega_0^{(i)}<\omega<\Gamma_i(p_{11}^{(i)})\\ \dfrac{(1-\varepsilon_i)(\beta p_{01}^{i)}+(1-\beta)\omega)}{1+\beta(p_{01}^{(i)}-\varepsilon_i p_{11}^{(i)})-\beta(1-\varepsilon_i)\omega}, & \text{if}\, \Gamma_i(p_{11}^{(i)})\leqslant\omega<\bar{\omega}_0^{(i)}\\ (1-\varepsilon_i)\omega, & \text{if}\, \bar{\omega}_0^{(i)}\leqslant\omega\leqslant p_{01}^{(i)}\end{cases} \tag{3.19}$$

$$W_\beta(\omega)=\begin{cases}(1-\varepsilon_i)\omega, & \text{if } p_{01}^{(i)}\leqslant\omega\leqslant\underline{\omega}_0^{1,(i)}\\ W_\beta\left(\underline{\omega}_0^{n,(i)}\right)+\left(\omega-\underline{\omega}_0^{n,(i)}\right)\dfrac{W_\beta\left(\overline{\omega}_0^{n(i)}\right)-W_\beta\left(\underline{\omega}_0^{n,(i)}\right)}{\overline{\omega}_0^{n,(i)}-\underline{\omega}_0^{n,(i)}}, & \text{if } \underline{\omega}_0^{n,(i)}<\omega<\overline{\omega}_0^{n,(i)}\, \text{and}\, n=1,2,\cdots\\ \dfrac{(1-\varepsilon_i)\left(1-\beta^{n+1}\right)(\omega-\beta\Gamma_i(\omega))+C_6}{C_0(\omega-\beta\Gamma_i(\omega))+C_7}, & \text{if } \overline{\omega}_0^{n,(i)}\leqslant\omega<\Gamma_i^n\left(\varphi\left(p_{11}^{(i)}\right)\right), n=1,2,\cdots\\ \dfrac{(1-\varepsilon_i)\left(1-\beta^{n+1}\right)(\omega-\beta\Gamma_i(\omega))+C_6}{(1-\varepsilon_i)\left(\beta-\beta^{n+1}\right)(\omega-\beta\Gamma_i(\omega))+C_9}, & \text{if } \Gamma_i^n\left(\varphi\left(p_{11}^{(i)}\right)\right)\leqslant\omega<\underline{\omega}_0^{n+1,(i)}, n=1,2,\cdots\\ \dfrac{(1-\varepsilon_i)\omega}{1-\beta(1-\varepsilon_i)\left(p_{11}^{(i)}-\omega\right)}, & \text{if } \omega_0^{(i)}\leqslant\omega\leqslant p_{11}^{(i)}\end{cases} \tag{3.20}$$

$$W(\omega)=\begin{cases}\dfrac{(1-\varepsilon_i)\left(p_{01}^{(i)}+\omega-\Gamma_i(\omega)\right)}{1+p_{01}^{(i)}-\varepsilon_i p_{11}^{(i)}-(1-\varepsilon_i)\Gamma_i\left(p_{11}^{(i)}\right)-(1-\varepsilon_i)(\omega-\Gamma_i(\omega))}, & \text{if } p_{11}^{(i)}\leqslant\omega<\omega_0^{(i)}\\ \dfrac{(1-\varepsilon_i)p_{01}^{(i)}}{1+p_{01}^{(i)}-\varepsilon_i p_{11}^{(i)}-(1-\varepsilon_i)\Gamma_i\left(p_{11}^{(i)}\right)}, & \text{if } \omega_0^{(i)}\leqslant\omega<\Gamma_i\left(p_{11}^{(i)}\right)\\ \dfrac{(1-\varepsilon_i)p_{01}^{(i)}}{1+p_{01}^{(i)}-\varepsilon_i p_{11}^{(i)}-(1-\varepsilon_i)\omega}, & \text{if } \Gamma_i\left(p_{11}^{(i)}\right)\leqslant\omega<\overline{\omega}_0^{(i)}\\ (1-\varepsilon_i)\omega, & \text{if } \overline{\omega}_0^{(i)}\leqslant\omega\leqslant p_{01}^{(i)}\end{cases} \tag{3.21}$$

$$W(\omega)=\begin{cases}(1-\varepsilon_i)\omega, & \text{if } p_{01}^{(i)}\leqslant\omega\leqslant\underline{\omega}_0^{1,(i)}\\ W\left(\underline{\omega}_0^{n,(i)}\right)+\left(\omega-\underline{\omega}_0^{n,(i)}\right)\dfrac{W\left(\overline{\omega}_0^{n,(i)}\right)-W\left(\underline{\omega}_0^{n,(i)}\right)}{\overline{\omega}_0^{n,(i)}-\underline{\omega}_0^{n,(i)}}, & \text{if } \underline{\omega}_0^{n,(i)}<\omega<\overline{\omega}_0^{n,(i)} \text{ and } n=1,2,\cdots\\ \dfrac{(1-\varepsilon_i)(n+1)(\Gamma_i(\omega)-\omega)-(1-\varepsilon)\Gamma_i^n\left(p_{01}^{(i)}\right)}{(1-\varepsilon_i)\left(n+1-(1-\varepsilon)p_{11}^{(i)}\right)(\Gamma_i(\omega)-\omega)+C_7'}, & \text{if } \overline{\omega}_0^{n,(i)}\leqslant\omega<\Gamma_i^n\left(\varphi\left(p_{11}^{(i)}\right)\right), n=1,2,\cdots\\ \dfrac{(1-\varepsilon)(n+1)(\Gamma_i(\omega)-\omega)-(1-\varepsilon)\Gamma_i^n\left(p_{01}^{(i)}\right)}{(1-\varepsilon)\left[n(\Gamma_i(\omega)-\omega)+p_{11}^{(i)}\right]-1-\Gamma_i^n\left(p_{01}^{(i)}\right)+\varepsilon\Gamma_i^n\left(p_{11}^{(i)}\right)}, & \text{if } \Gamma_i^n\left(\varphi\left(p_{11}^{(i)}\right)\right)\leqslant\omega<\underline{\omega}_0^{n+1,(i)}, n=1,2,\cdots\\ \dfrac{(1-\varepsilon_i)\omega}{1-(1-\varepsilon_i)\left(p_{11}^{(i)}-\omega\right)}, & \text{if } \omega_0^{(i)}\leqslant\omega\leqslant p_{11}^{(i)}\end{cases} \tag{3.22}$$

where,

$$C_0 =(1-\varepsilon_i)\beta[1-\beta^n p_{11}^{(i)}(1-\varepsilon_i)-\beta^{n+1}(1-(1-\varepsilon_i)p_{11}^{(i)})],$$

$$C_6 = (1-\varepsilon_i)(1-\beta)\beta^{n+1}\Gamma_i^n\left(p_{01}^{(i)}\right)$$

$$\begin{aligned}C_7 = &-\varepsilon_i(1-\beta)\beta^{n+1}\left(1-\beta(1-\varepsilon_i)p_{11}^{(i)}\right)\Gamma_i^n\left(p_{11}^{(i)}\right)\\ &+(1-\beta)\beta^{n+1}\left[1+\beta(1-\varepsilon_i)\left(1-p_{11}^{(i)}\right)\right]\Gamma_i^n\left(p_{01}^{(i)}\right)\\ &-\varepsilon_i(1-\varepsilon_i)(1-\beta)\beta^{n+1}p_{11}^{(i)}\Gamma_i^{n-1}\left(p_{11}^{(i)}\right)-(1-\varepsilon_i)(1-\beta)\beta^{n+1}\left(1-p_{11}^{(i)}\right)\Gamma_i^{n-1}\left(p_{01}^{(i)}\right)\\ &+(1-\beta)\left[1-\beta(1-\varepsilon_i)p_{11}^{(i)}\right]\end{aligned}$$

$$C_7' = \varepsilon_i\left[1-(1-\varepsilon_i)p_{11}^{(i)}\right]\Gamma_i^n\left(p_{11}^{(i)}\right) - \left[1+(1-\varepsilon_i)\left(1-p_{11}^{(i)}\right)\right]\Gamma_i^n\left(p_{01}^{(i)}\right)$$
$$+\varepsilon_i(1-\varepsilon_i)p_{11}^{(i)}\Gamma_i^{n-1}\left(p_{11}^{(i)}\right) + (1-\varepsilon_i)\left(1-p_{11}^{(i)}\right)\Gamma_i^{n-1}\left(p_{01}^{(i)}\right) + (1-\varepsilon_i)p_{11}^{(i)} - 1$$
$$C_9 = (1-\beta)\left[1-\beta(1-\varepsilon_i)p_{11}^{(i)}\right] + (1-\beta)\beta^{n+1}\left[\Gamma_i^n\left(p_{01}^{(i)}\right) - \varepsilon_i\Gamma_i^n\left(p_{11}^{(i)}\right)\right]$$

3.7 Fixed Point Analysis

In this section, we derive the fixed points of the mappings $ri(\cdot)$ and $Wi(\cdot)$ and their structural properties. To make our analysis concise, we omit the channel index i.

Lemma 3.1 (Fixed point of $\Gamma(\cdot)$) *Consider the case $p_{01} \leqslant p_{11}$, the following structural properties of* $\Gamma(\omega(t))$ *hold (see* Fig. 3.1):

(i) $\Gamma(\omega(t))$ *is monotonically increasing in* $\omega(t)$*;*

$$p_{01} \le \Gamma(\omega(t)) \le p_{11}, \forall 0 \le \omega(t) \le 1$$

(ii) $\Gamma^k(\omega(t)) = \Gamma(\Gamma^{k-1}(\omega(t)))$ *monotonically converges to* $\omega_0 := \frac{p_{01}}{1-(p_{11}-p_{01})}$ *as* $k \to \infty$.

Proof Noticing that $\Gamma(\omega(t))$ can be written as $\Gamma(\omega(t)) = (p_{11} - p_{01})\omega(t) + p_{01}$, Lemma 3.1 holds straightforwardly. □

Lemma 3.2 (Fixed point of $\Gamma(\cdot)$: $p_{01} > p_{11}$) *Denote* $\Gamma^0(\omega) = \omega$ *and* $\Gamma^k(\omega) = \Gamma(\Gamma^{k-1}(\omega))$*, then* $\Gamma^{2k}(\omega)$ *and* $\Gamma^{2k+1}(\omega)(\omega \in [p_{11}, p_{01}])$ *converge, from opposite directions, to* $\omega_0 := \frac{p_{01}}{1-(p_{11}-p_{01})}$ as $k \to \infty$ (see Fig. 3.2). *In particular, we have*

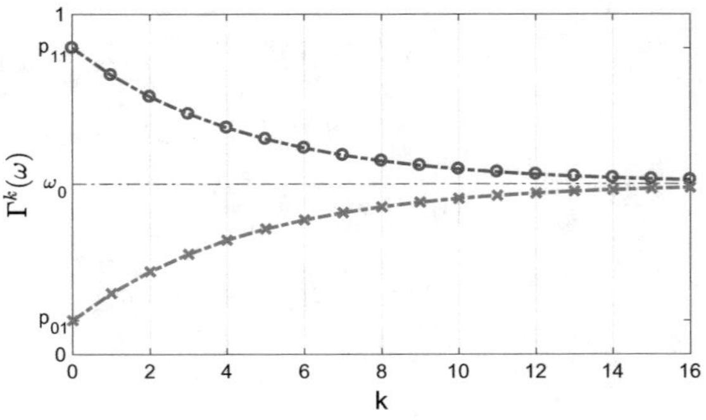

Fig. 3.1 $\Gamma^k(\omega)$ revolution as $k(p_{11} \geqslant p_{01})$. Red line: $\omega < \omega_0$; Green line: $\omega > \omega_0$

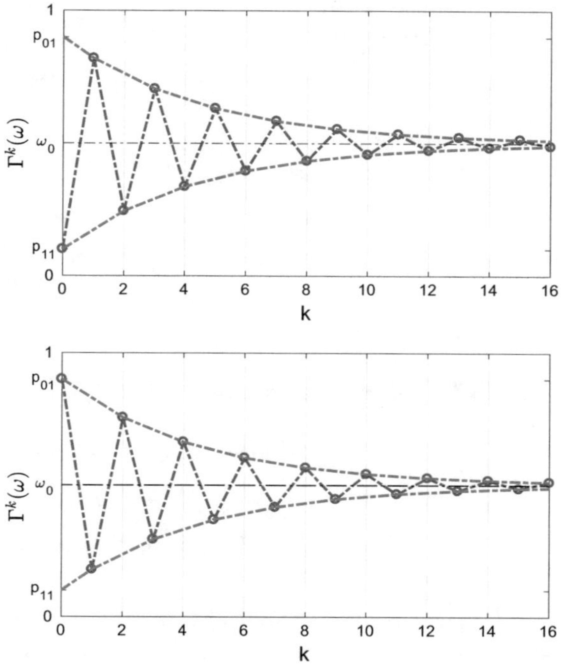

Fig. 3.2 $\Gamma^k(\omega)$ evolution as $k(p_{11} < p_{01})$. [Upper]: $p_{11} \leqslant \omega \leqslant \omega_0$; [Down]: $\omega_0 \leqslant \omega \leqslant p_{01}$—Red line indicates the envelop and green line indicates the evolution as k

$$\Gamma^k(\omega) > \omega \text{ if } p_{11} \leqslant \omega < \omega_0;$$

$$\Gamma^k(\omega_0) = \omega_0;$$

$$\Gamma^k(\omega) \leqslant \omega \text{ if } \omega_0 \leqslant \omega < p_{01}.$$

Proof It is easy to obtain the lemma, noticing $\Gamma(\omega) = (p_{11} - p_{01})\omega + p_{01}$ and $-1 < p_{11} - p_{01} < 0$.

Lemma 3.3 When $\varepsilon \leqslant \frac{(1-\max\{p_{11}, p_{01}\}) \cdot \min\{p, p_{01}\}}{(1-\min\{p_{11}, p_{01}\}) \cdot \max\{p_{11}, p_{01}\}}$, then

(i) *$\varphi(\omega(t))$ monotonically increases with $\omega(t)$;*

$$\varphi(\omega(t)) \leqslant \min\{p_{11}, p_{01}\}, \min\{p_{11}, p_{01}\} \leqslant \omega(t) \leqslant \max\{p_{11}, p_{01}\}$$

$$\varphi(0) = 0, \varphi(1) = 1$$

□

Proof According to (3.3) and (3.4), it is easy to obtain the results. ■

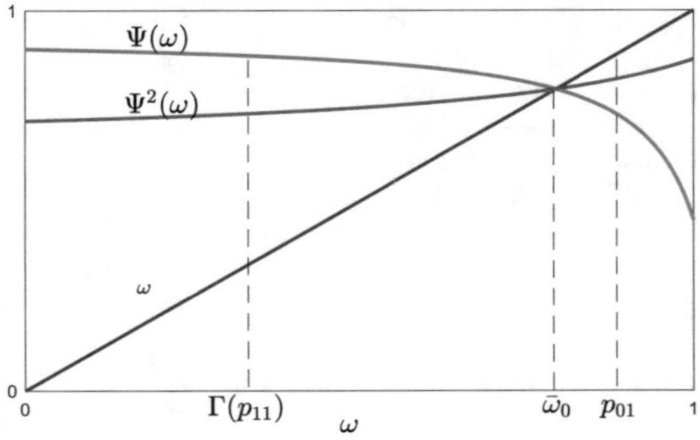

Fig. 3.3 Functions $\Psi^2(\omega)$, $\Psi(\omega)$, ω with $\omega(p_{11} < p_{01})$

Lemma 3.4 *Given* $p_{01} > p_{11}$, *there exists* $\omega_0 \in [\Gamma(p_{11}), p_{01}]$ (see Fig. 3.3) *such that*

$$\Psi(\omega) > \omega, \text{if } \Gamma(p_{11}) \leqslant \omega < \overline{\omega}_0$$

$$\Psi(\omega_0) = \omega_0$$

$$\Psi(\omega) < \omega, \text{if } \overline{\omega}_0 \leqslant \omega < p_{01}$$

Proof Since $\varphi(\omega)$ monotonically increases with ω while $\Gamma(\omega)$ monotonically decreases with ω when $p_{11} < p_{01}$, we obtain that $\Psi(\omega) = \Gamma(\varphi(\omega))$ decreases monotonically with ω, and moreover, $\varphi(\omega)$ is concave in ω since the following condition:

$$\begin{cases} \dfrac{\partial[\Psi(\omega)]}{\partial[\omega]} = -\dfrac{\varepsilon(p_{01} - p_{11})}{[1-(1-\varepsilon)\omega]^2} < 0 \\ \dfrac{\partial^2 |\Psi(\omega)]}{\partial^2 |\omega]} = -\dfrac{2\varepsilon(1-\varepsilon)(p_{01} - p_{11})}{[1-(1-\varepsilon)\omega]^3} < 0. \end{cases}$$

Next, we show that there exists $\overline{\omega}_0 \in [\Gamma(p_{11}), p_{01}]$ such that $\Psi(\overline{\omega}_0) = \overline{\omega}_0$. Since the following inequalities hold regarding two endpoints $\Gamma(p_{11})$ and p_{01}

$$\begin{cases} \Psi(\Gamma(p_{11})) = \Gamma(\varphi(\Gamma(p_{11}))) > \Gamma(p_{11}) \\ \Psi(p_{01}) = \Gamma(\varphi(p_{01})) < \Gamma(\varphi(0)) = \Gamma(0) = p_{01} \end{cases}$$

we know that $\Psi(\omega)$ and ω must have a unique intersection at some point $\overline{\omega}_0$ as shown in Fig. 3.6, taking into account the concavity of $\Psi(\omega)$ and the linearity of to, which indicates that $\Psi(\omega) > \omega_0 > \omega$ when $\Gamma(p_{11}) \leqslant \omega < \omega_0$ while $\Psi(\omega) \leqslant \omega_0 \leqslant \omega$ when $\overline{\omega}_0 \leqslant \omega < p_{01}$ □

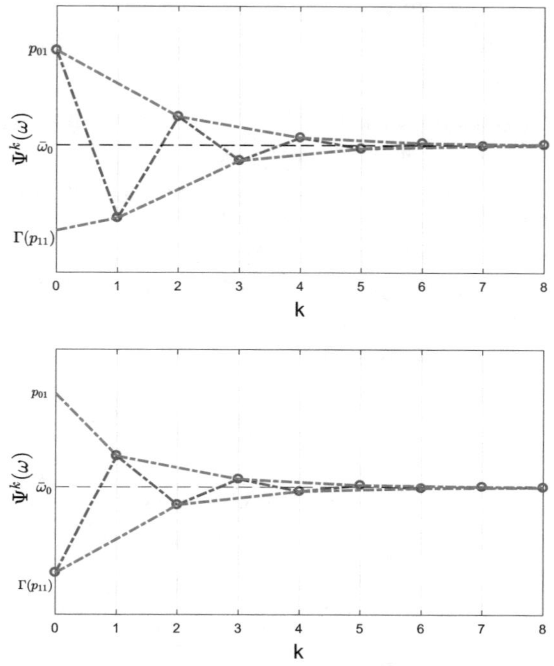

Fig. 3.4 $\Psi^k(\omega)$ evolution as $k(p_{11} < p_{01})$. [U]: $\omega_0 \leqslant \omega \leqslant p_{01}$; [D]: $\Gamma(p_{11}) \leqslant \omega \leqslant \omega_0$ Red line indicates the envelop and greed line indicates the evolution as k

Lemma 3.5 (Fixed point of $\Psi(\omega)$: $p_{01} > p_{11}$) *Let* $\Psi^0(\omega) = \omega$ and $\Psi^k(\omega) = \Psi(\Psi^{k-1}(\omega))$, $\Psi^{2k}(\omega)$ *and* $\Psi^{2k+1}(\omega)(\omega \in [\Gamma(p_{11}), p_{01}])$ *converge, from opposite directions, to* $\overline{\omega}_0$ *as* $k \to \infty$ (see Fig. 3.4). *In particular, we have*

$$\Psi^k(\omega) \leqslant \omega \text{ if } \overline{\omega}_0 \leqslant \omega < p_{01};$$

$$\Psi^k(\overline{\omega}_0) = \omega_0;$$

$$\Psi^k(\omega) > \omega \text{ if } \Gamma(p_{11}) \leqslant \omega < \omega_0.$$

Proof We prove the lemma in two different cases.

Case 1. When $\omega_0 \leqslant \omega < p_{01}$, to show $\Psi^i(\omega) \leqslant \omega$, it is sufficient to prove

$$\begin{cases} \omega \geqslant \Psi^0(\omega) > \cdots > \Psi^{2k}(\omega) > \Psi^{2k+2}(\omega) > \cdots > \overline{\omega}_0 \\ \Psi^1(\omega) < \cdots < \Psi^{2k+1}(\omega) < \Psi^{2k+3}(\omega) < \cdots < \overline{\omega}_0 \leqslant \omega. \end{cases} \tag{3.23}$$

To prove (3.23), it is sufficient to show for $k = 0, 1, 2, \cdots$

(i) when $\Psi^{2k}(\omega) > \overline{\omega}_0, \Psi^{2k+2}(\omega) > \overline{\omega}_0$;
(ii) when $\Psi^{2k+1}(\omega) < \omega_0$, $\Psi^{2k+3}(\omega) < \omega_0$.

(iii) when $\Psi^{2k}(\omega) > \omega_0$, $\Psi^{2k}(\omega) > \Psi^{2k+2}(\omega)$;
(iv) when $\Psi^{2k+1}(\omega) < \overline{\omega}_0, \Psi^{2k+1}(\omega) < \Psi^{2k+3}(\omega)$

First, we prove (i). By Lemma 3.4, $\Psi^0(\omega) \geqslant \omega_0 \geqslant \Psi^1(\omega)$ for $k = 0$. When $\Psi^{2k}(\omega) > \omega_0$, we have $\Psi^{2k+1}(\omega) = \Psi(\Psi^{2k}(\omega)) < \omega_0$ by Lemma 3.4, and furthermore, $\Psi^{2k+2}(\omega) = \Psi\left(\Psi^{2k+1}(\omega)\right) > \overline{\omega}_0$.

Second, we prove (ii). By Lemma 3.4, $\Psi^0(\omega) \geqslant \overline{\omega}_0 \geqslant \Psi^1(\omega)$ for $k = 0$. When $\Psi^{2k+1}(\omega) < \omega_0$, we have $\Psi^{2k+2}(\omega) = \Psi(\Psi^{2k+1}(\omega)) > \omega_0$ by Lemma 3.4, and furthermore, $\Psi^{2k+3}(\omega) = \Psi\left(\Psi^{2k+2}(\omega)\right) < \overline{\omega}_0$.

Third, to prove (iii) and (vi), we only need to show that $\Psi^2(\omega) < \omega$ for any $\omega \in (\overline{\omega}_0, p_{01}]$ while $\Psi^2(\omega) > \omega$ for any $\omega \in [\Gamma(p_{11}), \overline{\omega}_0)$.

On one hand, from the following:

$$\frac{\partial[\Psi^2(\omega)]}{\partial[\omega]} = \frac{\partial[\Psi(x)]}{\partial[x]}\bigg|_{x=\Psi(\omega)} \cdot \frac{\partial[\Psi(\omega)]}{\partial[\omega]}$$

$$= \frac{\varepsilon(p_{01} - p_{11})}{[1-(1-\varepsilon)\Psi(\omega)]^2} \cdot \frac{\varepsilon(p_{01} - p_{11})}{[1-(1-\varepsilon)\omega]^2}$$

$$= \frac{\varepsilon^2(p_{01} - p_{11})^2}{[1-(1-\varepsilon)((1-p_{01}+\varepsilon p_{11})\omega + p_{01})]^2} > 0 \tag{3.24}$$

$$\frac{\partial^2[\Psi^2(\omega)]}{\partial^2[\omega]} = \frac{2(1-\varepsilon)\varepsilon^2(p_{01}-p_{11})^2(1-p_{01}+\varepsilon p_{11})}{[1-(1-\varepsilon)((1-p_{01}+\varepsilon p_{11})\omega + p_{01})]^3} > 0, \tag{3.25}$$

we have $\Psi^2(\omega)$ is the convex increasing function of ®.

On the other hand, we have the following inequalities regarding three end points:

$$\begin{cases} \Psi^2(\Gamma(p_{11})) = \Psi(\Psi(\Gamma(p_{11}))) > \Psi(\Gamma(p_{11})) > \Gamma(p_{11}) \\ \Psi^2(\overline{\omega}_0) = \Psi(\Psi(\overline{\omega}_0)) = \Psi(\overline{\omega}_0) = \overline{\omega}_0 \\ \Psi^2(p_{01}) = \Psi(\Psi(p_{01})) < \Psi(0) = p_{01} \end{cases} \tag{3.26}$$

Therefore, combining (3.24), (3.25), and (3.26), we know that $\Psi^2(\omega)$ and ω have a unique intersection at $\overline{\omega}_0$ as shown in Fig. 3.3, and furthermore, $\Psi^2(\omega) < \omega$ for any $\omega \in (\omega_0, p_{01}]$ while $\Psi^2(\omega) > \omega$ for any $\omega \in [\Gamma(p_{11}), \omega_0)$.

Case 2. When $\Gamma(p_{11}) \leqslant \omega < \overline{\omega}_0$, we need to prove

$$\begin{cases} \omega = \Psi^0(\omega) < \cdots < \Psi^{2k}(\omega) < \Psi^{2k+2}(\omega) < \cdots < \omega_0 \\ \Psi(\omega) > \cdots > \Psi^{2k+1}(\omega) > \Psi^{2k+3}(\omega) > \cdots > \omega_0 > \omega \end{cases}$$

which can be verified by the similar induction in the aforementioned case. Therefore, combining the above two cases, we conclude the lemma. □

3.8 Threshold Policy and Adjoint Dynamic System

In this section, we first express the value function by threshold policy, and then introduce an adjoint dynamic system to facilitate the analysis on nonlinear dynamics.

3.8.1 Threshold Policy

Let $L(\omega, \omega')$ be the minimum amount of time required for a passive arm to transit across ω' starting from ω, i.e.,

$$L(\omega, \omega') \triangleq \min\left\{k : \Gamma^k(\omega) > \omega'\right\} \tag{3.27}$$

According to Lemma 3.1, we have for the case of $p_{11} \geqslant p_{01}$

$$L(\omega, \omega') = \begin{cases} 0 & \text{if } \omega > \omega' \\ \lfloor \log_{p_{11}-p_{01}} \frac{\omega_0 - \omega'}{\omega_0 - \omega} \rfloor + 1, & \text{if } \omega \leqslant \omega' < \omega_0 \\ \infty & \text{if } \omega \leqslant \omega', \omega' \geqslant \omega_0 \end{cases} \tag{3.28}$$

and, for the case of $p_{11} < p_{01}$

$$L(\omega, \omega') = \begin{cases} 0, & \text{if } \omega > \omega' \\ 1, & \text{if } \omega \leqslant \omega' \text{ and } \Gamma(\omega) > \omega' \\ \infty, & \text{if } \omega \leqslant \omega' \text{ and } \Gamma(\omega) \leqslant \omega' \end{cases} \tag{3.29}$$

Under the threshold policy, the arm will be activated if its belief state crosses a certain threshold ω'. In other words, starting from an arbitrary belief state «, the first active action on the arm is taken after $L(\omega, \omega')$ slots.

Based on the structure of threshold policy, $V_{\beta,\, v}(\omega)$ can be characterized in terms of $V_{\beta,v}\left(\Gamma^{t_0-1}(\omega); a = 1\right)$ for some $t_0 \in \{1, 2, \cdots, \infty\}$ where $t_0 = L(\omega, \omega^*) + 1$ is the slot when the belief ω reaches the threshold ω^* for the first time. Specially, in the first $L(\omega, \omega^*)$ slots, the subsidy v is obtained in each slot. In slot $t_0 = L(\omega, \omega^*) + 1$, the belief state reaches the threshold or and the arm is activated. The total reward thereafter is $V_{\beta,v}\left(\Gamma^{L(\omega,\omega^*)}(\omega); a = 1\right)$. Taking into account β, we thus have

$$V_{\beta,v}(\omega) = \frac{1 - \beta^{L(\omega,\omega^*)}}{1 - \beta} v + \beta^{L(\omega,\omega^*)} V_{\beta,v}\left(\Gamma^{L(\omega,\omega^*)}(\omega); a = 1\right) \tag{3.30}$$

3.8.2 Adjoint Dynamic System

In the dynamic system (3.15), the belief information ω represents two kinds of information:

- policy information, i.e., action a depends on ω;
- value information, i.e., the reward value of the dynamic system (or value function) depends on ω.

To better characterize the dynamic evolution of (3.15), we separate two roles of ω by mathematically letting ω only represent the value while introducing $\lfloor \omega \rceil$ to indicate the policy information used to make a decision (corresponding to the policy).

Specifically, we introduce the following adjoint dynamic system:

$$V_{\beta,v}(\omega;\omega') = \max\left\{V_{\beta,v}(\omega;\omega',0), V_{\beta,v}(\omega;\omega',1)\right\} \tag{3.31}$$

where,

$$\begin{aligned} V_{\beta,v}(\omega;\omega',0) &= v + \beta V_{\beta,v}(\Gamma(\omega);\Gamma(\omega')), \\ V_{\beta,v}(\omega;\omega',1) &= (1-\varepsilon)\omega + \beta\big[(1-\varepsilon)\omega V_{\beta,v}(p_{11};p_{11}) \\ &\quad +(1-(1-\varepsilon)\omega)V_{\beta,v}(\Psi(\omega);\Psi(\omega'))\big] \end{aligned}$$

where ω', a represents making action $a(a = 0{,}1)$ given the policy information ω'.

Proposition 3.1 *Given* v, $V_{\beta,v}(\omega; a = 1)$ *and* $V_{\beta,v}(\omega; a = 0)$ *are piecewise linear and convex in* ω.

Proof We prove the proposition by induction. In slot T, we have $V^T_{\beta,v}(\omega; a = 0) = v$ and $V^T_{\beta,v}(\omega; a = 1) = (1-\varepsilon)\omega$, which follows $V^T_{\beta,v}(\omega) = \max\{V^T_{\beta,v}(\omega; a = 0), V^T_{\beta,v}(\omega, a = 1)\}$ is piecewise linear and convex in ω.

Assume $V^{t+1}_{\beta,v}(\omega; a = 1)$ and $V^{t+1}_{\beta,v}(\omega; a = 0)$ are piecewise linear and convex in ω, it is easy to show that both $V^t_{\beta,v}(\omega; a = 1)$ and $V^t_{\beta,v}(\omega; a = 0)$ are piecewise linear and convex in ω according to Eq. (3.15). Letting $T \nearrow \infty$, we prove the proposition. □

Corollary 3.2 $V^t_{\beta,v}(\omega;\omega')$ *is piecewise linear in* ω.

Proof We know that $V^t_{\beta,v}(\omega;\omega)$ is piece linear in ω by Proposition 3.1. In $V^t_{\beta,v}(\omega;\omega')$, ω represents the value information while ω' does the policy information. Thus, $V^t_{\beta,v}(\omega;\omega')$ is piece linear in the value information ω. □

Lemma 3.6 *$V_{\beta,v}(\omega;\omega',1)$ is decomposable in ω, i.e.,*

$$V_{\beta,v}(\omega;\omega',1) = (1-\varepsilon)\omega + \beta\big[(1-\varepsilon)\omega V_{\beta,v}(p_{11};p_{11}) \\ +\varepsilon\omega V_{\beta,v}(p_{11};\Psi(\omega')) + (1-\omega)V_{\beta,v}(p_{01};\Psi(\omega'))\big].$$

Proof

$$\begin{aligned} V_{\beta,v}(\omega;\omega',1) &= (1-\varepsilon)\omega + \beta\big[(1-\varepsilon)\omega V_{\beta,v}(p_{11};p_{11}) \\ &\quad +(1-(1-\varepsilon)\omega)V_{\beta,v}(\Psi(\omega);\Psi(\omega'))\big] \\ &\overset{(a)}{=} (1-\varepsilon)\omega + \beta\big[(1-\varepsilon)\omega V_{\beta,v}(p_{11};p_{11}]) \\ &\quad +(1-\omega)(1-(1-\varepsilon)0)V_{\beta,v}(\Psi(0);\Psi(\omega')) \\ &\quad +\omega(1-(1-\varepsilon)1)V_{\beta,v}(\Psi(1);\Psi(\omega'))\big] \\ &= (1-\varepsilon)\omega + \beta\big[(1-\varepsilon)\omega V_{\beta,v}(p_{11};p_{11}) \\ &\quad +(1-\omega)V_{\beta,v}(p_{01};\Psi(\omega')) + \varepsilon\omega V_{\beta,v}(p_{11};\Psi(\omega'))\big] \end{aligned}$$

where (a) is due to Corollary 3.2. □

Remark 3.3 In (3.32), for $V_{\beta,v}(p_{11};p_{11})$ and $V_{\beta,v}(p_{11};\Psi(\omega'))$, we can see that though they have the same value information p_{11}, they have different policy information, i.e., p_{11} and $\Psi(\omega')$ Hence, $V_{\beta,v}(p_{11};p_{11}) \neq V_{\beta,v}(p_{11};\Psi(\omega'))$ except that both p_{11} and $\Psi(\omega')$ can lead a same action policy for the dynamic system.

3.9 Linearization of Value Function for Negatively Correlated Channels

In this section, we focus on the linearization of $V_{\beta,v}(\omega;\omega,1)$ for the case of negatively correlated channels, i.e., $p_{11}^{(i)} < p_{01}^{(i)}$, which serves as the basis to compute the Whittle index. Again, we consider one channel by dropping channel index i.

In many practical systems, the initial belief ω_0 is set to ω_0[9]. It can then be checked that $\min\{p_{01},p_{11}\} \leqslant \omega_0 \leqslant \max\{p_{01},p_{11}\}$. Moreover, even the initial belief does not fall in $[\min\{p_{01},p_{11}\},\max\{p_{01},p_{11}\}]$, all the belief values are bounded in the interval from the second slot following Lemma 3.1. Hence the following results can be extended by treating the first slot separately from the future slots. Therefore, we assume $\min\{p_{01},p_{11}\} \leqslant \omega \leqslant \max\{p_{01},p_{11}\}$ in the first slot in our analysis.

We divide the region $[p_{11},p_{01}]$ into four subregions by $\Gamma(p_{11})$ and two fixed points $\omega_0,\overline{\omega}_0$:

$$[p_{11},p_{01}] = [p_{11},\omega_0) \cup [\omega_0,\Gamma(p_{11})) \cup [\Gamma(p_{11}),\omega_0) \cup [\omega_0,p_{01}] \tag{3.33}$$

In the following, we derive the linearized $V_{\beta,\ v}(\omega;\omega,1)$ in these subregions, respectively, which will be used to compute the Whittle index in Sect. 3.12.

3.9.1 Region 1–2

Proposition 3.2 *If $p_{11} \leqslant \omega^* < \Gamma(p_{11})$, it holds that $L(\Gamma(\varphi(\omega)),\omega^*) = 0$ for any $\omega \in [p_{11}, p_{01}]$*

Proof In the case of $p_{11} < p_{01}$, $\varphi(\omega$ monotonically increases with ω while $\Gamma(\omega)$ monotonically decreases with ω. Thus, $\Gamma(\varphi(\omega)) \geqslant \Gamma(p_{11}) > \omega^*$ for $\omega \in [p_{11}, p_{01}]$ when $0 \leqslant \varepsilon \leqslant \frac{p_{11}(1-p_{01})}{p_{01}(1-p_{11})}$. Therefore, $L(\Gamma(\varphi(\omega)),\omega^*) = 0$. □

Lemma 3.7 *When $p_{11} \leqslant \omega^* < \Gamma(p_{11})$ for any $\omega \in [p_{11}, p_{01}]$, the linearity of $V_{\beta,v}(\omega;\omega,1)$ is as follows:*

$$\begin{aligned} V_{\beta,v}(\omega;\omega,1) = (1-\varepsilon)\omega + \beta\big[(1-\varepsilon)\omega V_{\beta,v}(p_{11};p_{11}) \\ +\varepsilon\omega V_{\beta,v}(p_{11};\Psi(\omega)) + (1-\omega)V_{\beta,v}(p_{01};\Psi(\omega))\big] \end{aligned} \tag{3.34}$$

where

$$\begin{aligned} V_{\beta,v}(p_{11};p_{11}) &\overset{(e1)}{=} V_{\beta,v}(p_{11};p_{11},0) \\ &\overset{(e2)}{=} v + \beta V_{\beta,v}(\Gamma(p_{11});\Gamma(p_{11})) \\ &\overset{(e3)}{=} v + \beta(1-\varepsilon)\Gamma(p_{11}) + \beta^2\big[(1-\varepsilon)\Gamma(p_{11})V_{\beta,v}(p_{11};p_{11}) \\ &\quad +(1-(1-\varepsilon)\Gamma(p_{11}))V_{\beta,v}(\Psi(\Gamma(p_{11}));\Psi(\Gamma(p_{11})))\big] \\ &\overset{(e4)}{=} v + \beta(1-\varepsilon)\Gamma(p_{11}) + \beta^2\big[(1-\varepsilon)\Gamma(p_{11})V_{\beta,v}(p_{11};p_{11}) \\ &\quad +(1-(1-\varepsilon)\Gamma(p_{11}))V_{B,v}(\Psi(\Gamma(p_{11}));\Psi(\omega))\big] \\ &\overset{(e5)}{=} v + \beta(1-\varepsilon)\Gamma(p_{11}) \\ &\quad + \beta^2[(1-\varepsilon)\Gamma(p_{11})V_{\beta,v}(p_{11};p_{11}) \\ &\quad + \varepsilon\Gamma(p_{11})V_{\beta,v}(p_{11};\Psi(\omega)) \\ &\quad +(1-\Gamma(p_{11}))V_{\beta,v}(p_{01};\Psi(\omega))] \end{aligned} \tag{3.35}$$

$$\begin{aligned} V_{\beta,v}(p_{11};\Psi(\omega)) &= V_{\beta,v}(p_{11};\Psi(\omega),1) \\ &\overset{(e6)}{=} (1-\varepsilon)p_{11} + \beta\big[(1-\varepsilon)p_{11}V_{\beta,v}(p_{11};p_{11}) \\ &\quad +\varepsilon p_{11}V_{\beta,v}(p_{11};\Psi(\omega)) + (1-p_{11})V_{\beta,v}(p_{01};\Psi(\omega))\big] \end{aligned} \tag{3.36}$$

$$\begin{aligned}
V_{\beta,v}(p_{01};\Psi(\omega)) &= V_{\beta,v}(p_{01};\Psi(\omega),1) \\
&\overset{(e7)}{=} (1-\varepsilon)p_{01} + \beta\big[(1-\varepsilon)p_{01}V_{\beta,v}(p_{11};p_{11}) \\
&+\varepsilon p_{01}V_{\beta,v}(p_{11};\Psi(\omega)) + (1-p_{01})V_{\beta,v}(p_{01};\Psi(\omega))\big].
\end{aligned} \tag{3.37}$$

Proof (e1) is due to $p_{11} \le \omega^* \Rightarrow a = 0$ (e2) is due to $a = 0$. (e3) is due to $\Gamma(p_{11}) > \omega^* \Rightarrow a = 1$. (e4) is due to $L(\Psi(\Gamma(p_{11})), \omega^*) = 0$ and $L(\Psi(\omega), \omega^*) = 0$ from Proposition 3.2. (e5), (e6), and (e7) follow Lemma 3.6. □

Remark 3.4 Based on (3.35), (3.36), and (3.37), we can compute $V_{\beta,v}(p_{11};p_{11})$,$V_{\beta,v}(p_{11};\Psi(\omega))$, and $V_{\beta,v}(p_{01};\Psi(\omega))$, and further $V_{\beta,v}(\omega;\omega,1)$ is linearized by (3.34).

3.9.2 Region 3

Based on Lemma 3.5, we have the following important corollary.

Corollary 3.3 *When $\Gamma(p_{11}) \leqslant \omega^* < p_{01}$, we have*

(i) When $\Gamma(p_{11}) \leqslant \omega^* < \overline{\omega}_0$ the first crossing time of the nonlinear belief part $\Psi^i(\omega^*)$ $(i = 1, 2, \cdots)$ will be 0 in the evolving process; that is, $L(\Psi^i(\omega^*), \omega^*) = 0$;
(ii) When $\overline{\omega}_0 \leqslant \omega^* < p_{01}$ first crossing time of the nonlinear belief part $\Gamma^i(\Psi(\omega^*))$ $(i = 0, 1, 2, \cdots)$ will be ∞; that is, $L(\Gamma^i(\Psi(\omega^*)), \omega^*) = \infty$.

Proof (1) By Lemma 3.5, we have that $\Psi^i(\omega^*) > \omega^*$ when $\Gamma(p_{11}) \leqslant \omega^* < \omega_0$, and furthermore, $L(\Psi^i(\omega^*), \omega^*) = 0$. (2) By Lemma 3.5, $\omega_0 < \Psi(\omega^*) \leqslant \omega^*$ when $\overline{\omega}_0 \leqslant \omega^* < p_{01}$. Furthermore, by Lemma 3.2, we have $\Gamma^i(\Psi'(\omega^*)) \leqslant \omega^*$, which means $L(\Gamma^i(\Psi(\omega^*)), \omega^*) = \infty$ □

Corollary 3.4 *When $\Gamma(p_{11}) \leqslant \omega^* < \overline{\omega}_0$, we have that $V_{\beta,v}(\omega^*;\omega^*,1)$ can be linearized by the following:*

$$\begin{aligned}
V_{\beta,v}(\omega^*;\omega^*,1) &\overset{(e1)}{=} (1-\varepsilon)\omega^* + \beta\big[(1-\varepsilon)\omega^* V_{\beta,v}(p_{11},p_{11}) \\
&\quad +(1-(1-\varepsilon)\omega^*)V_{\beta,v}(\Psi(\omega^*);\Psi(\omega^*))\big] \\
&\overset{(e2)}{=} (1-\varepsilon)\omega^* + \beta\big[(1-\varepsilon)\omega^* V_{\beta,v}(p_{11};p_{11}) \\
&\quad +\varepsilon\omega^* V_{\beta,v}(p_{11};\Psi(\omega^*)) + (1-\omega^*)V_{\beta,v}(p_{01};\Psi(\omega^*))\big]
\end{aligned}$$

where

$$V_{\beta,v}(p_{11};p_{11}) \overset{(e3)}{=} \frac{v}{1-\beta},$$

$$\begin{aligned}
V_{\beta,v}(p_{11};\Psi(\omega^*)) &\overset{(e4)}{=} V_{\beta,v}(p_{11};\Psi(\omega^*),1)\\
&\overset{(e5)}{=} (1-\varepsilon)p_{11} + \beta\big[(1-\varepsilon)p_{11}V_{\beta,v}(p_{11};p_{11})\\
&\quad +(1-(1-\varepsilon)p_{11})V_{\beta,v}\big(\Psi(p_{11});\Psi^2(\omega^*)\big)\big]\\
&\overset{(e6)}{=} (1-\varepsilon)p_{11} + \beta\big[(1-\varepsilon)p_{11}V_{\beta,v}(p_{11};p_{11})\\
&\quad +(1-(1-\varepsilon)p_{11})V_{\beta,v}(\Psi(p_{11});\Psi(\omega^*))\big]\\
&\overset{(e7)}{=} (1-\varepsilon)p_{11} + \beta\big[(1-\varepsilon)p_{11}V_{\beta,v}(p_{11};p_{11})\\
&\quad +\varepsilon p_{11}V_{\beta,v}(p_{11};\Psi(\omega^*))\\
&\quad +(1-p_{11})V_{\beta,v}(p_{01};\Psi(\omega^*))\big]
\end{aligned}$$

$$\begin{aligned}
V_{\beta,v}(p_{01};\Psi(\omega^*)) &\overset{(e8)}{=} (1-\varepsilon)p_{01} + \beta\big[(1-\varepsilon)p_{01}V_{\beta,v}(p_{11};p_{11})\\
&\quad +\varepsilon p_{01}V_{\beta,v}(p_{11};\Psi(\omega^*))\\
&\quad +(1-p_{01})V_{\beta,v}(p_{01};\Psi(\omega^*))\big].
\end{aligned}$$

Proof (e1) is due to $a = 1$. (e2) is due to Lemma 3.6. (e3) is due to $L(p_{11},\omega^*) = \infty \Rightarrow a = 0$ for $\Gamma(p_{11}) \leqslant \omega^* < \overline{\omega}_0$ and $V_{\beta,v}(p_{11};p_{11}) = v + \beta v + \beta^2 v^2 + \cdots = \frac{v}{1-B}$ is due to $L(\Psi(\omega^*),\omega^*) = 0 \Rightarrow a = 1$ from Corollary 3.3. (e6) is due to $L(\Psi^2(\omega^*),\omega^*) = L(\Psi'(\omega^*),\omega^*)$.l(e7) is due to Corollary 3.2.(e8) is due to $a = 1 \Leftarrow L(\Psi(\omega^*),\omega^*) = 0$ by Corollary 3.3. □

3.9.3 Region 4

When $\omega_0 \leqslant \omega^* < p_{01}$, we have $L(\Gamma^i(\Psi(\omega^*)),\omega^*) = \infty$ by Corollary 3.3, and further $V_{\beta,v}(\Psi(\omega^*);\Psi(\omega^*)) = \frac{v}{1-\beta}$. Hence,

$$\begin{aligned}
V_{\beta,v}(\omega^*;\omega^*,1) &= (1-\varepsilon)\omega^* + \beta\big[(1-\varepsilon)\omega^* V_{\beta,v}(p_{11};p_{11})\\
&\quad +(1-(1-\varepsilon)\omega^*)V_{\beta,v}(\Psi(\omega^*);\Psi(\omega^*))\big]
\end{aligned}$$

$$= (1-\varepsilon)\omega^* + \beta\left[(1-\varepsilon)\omega^* V_{\beta,v}(p_{11};p_{11}) + (1-(1-\varepsilon)\omega^*)\frac{v}{1-\beta}\right]$$

which clearly shows that $V_{\beta,v}(\omega^*;\omega^*,1)$ has been linearized.

3.10 Linearization of Value Function for Positively Correlated Channels

In this section, we linearize $V_{\beta,\nu}(\omega;\omega,1)$ for the case of $p_{11} \geqslant p_{01}$.

According to the fixed point ω_0, we first divide $[p_{01}, p_{11}]$ into $[p_{01}, \omega_0)$ and $[\omega_0, p_{11}]$. According to Lemmas 3.1 and 3.3, $[p_{01}, \omega_0)$ can be further divided into $[p_{01}, \Gamma(\varphi(p_{01})))$ and $[\Gamma^n(\varphi(p_{01})), \Gamma^{n+1}(\varphi(p_{01})))$, herein, $n = 1, 2, \cdots, \infty$. Thus, in the following, we take $[\Gamma^n(\varphi(p_{01})), \Gamma^{n+1}(\varphi(p_{01})))$ as an example to analyze the evolution of belief value.

Lemma 3.8 (Fixed points) *If* $p_{11} \geqslant p_{01}$, *for any* $n(n = 1, 2, \cdots, \infty)$ *there exists* $\underline{\omega}_0^n$, *and* $\overline{\omega}_0^n$ *satisfying* $\Gamma^n(\varphi(p_{01})) < \underline{\omega}_0^n < \overline{\omega}_0^n < \Gamma^n(\varphi(p_{11}))$ *such that*

$$\varphi(\Gamma(\omega)) > \Gamma^{-n}(\omega) \text{ for } p_{01} \leqslant \omega < \overline{\omega}_0^n$$
$$\varphi(\Gamma(\overline{\omega}_0^n)) = \Gamma^{-n}(\overline{\omega}_0^n)$$
$$\varphi(\Gamma(\omega)) < \Gamma^{-n}(\omega) \text{ for } \overline{\omega}_0^n < \omega \leqslant \omega_0$$

and

$$\varphi(\omega) > \Gamma^{-n}(\omega) \text{ for } p_{01} \leqslant \omega < \underline{\omega}_0^n$$
$$\varphi(\underline{\omega}_0^n) = \Gamma^{-n}(\underline{\omega}_0^n)$$
$$\varphi(\omega) < \Gamma^{-n}(\omega) \text{ for } \underline{\omega}_0^n < \omega \leqslant \omega_0$$

Proof Since

$$\frac{\partial[\varphi(\Gamma(\omega))]}{\partial[\omega]} = \frac{\varepsilon(p_{11} - p_{01})}{[1 - (1-\varepsilon)\omega]^2} > 0$$
$$\frac{\partial^2[\varphi(\Gamma(\omega))]}{\partial^2[\omega]} = \frac{2\varepsilon(1-\varepsilon)(p_{11} - p_{01})}{[1 - (1-\varepsilon)\omega]^3} > 0,$$

we know that $\varphi(\Gamma(\omega))$ is convex in ω. Moreover, by the increasing monotonicity of $\varphi(\omega)$, we have the following inequalities regarding end points $\Gamma^n(\varphi(p_{01}))$ and $\Gamma^n(\varphi(p_{11}))$

$$\begin{cases} \varphi(\Gamma(\Gamma^n(\varphi(p_{01})))) > \varphi(p_{01}) = \Gamma^{-n}(\Gamma^n(\varphi(p_{01}))) \\ \varphi(\Gamma(\Gamma^n(\varphi(p_{11})))) < \varphi(p_{11}) = \Gamma^{-n}(\Gamma^n(\varphi(p_{11}))) \end{cases}$$

Combining the linearity of $\Gamma^{-n}(\omega) = (p_{11} - p_{01})\omega_0 + \frac{\omega - (p_{11} - p_{01})\omega_0}{(p_{11} - p_{01})^n}$ and the convexity of $\varphi(\Gamma(\omega))$ in ω, we know that there must exist a unique point

$\overline{\omega}_0^n(\Gamma^n(\varphi(p_{01})) < \overline{\omega}_0^n < \Gamma^n(\varphi(p_{11})))$ such that $\varphi(\Gamma(\omega)) \leqslant \Gamma^{-n}(\omega)$ when $\overline{\omega}_0^n \leqslant \omega \leqslant \omega_0$ while $\varphi(\Gamma(\omega)) > \Gamma^{-n}(\omega)$ when $p_{01} \leqslant \omega < \overline{\omega}_0^n$.

Likewise, there exists a unique point $\omega_l^n\left(\Gamma^n(\varphi(p_{01})) < \underline{\omega}_0^n < \Gamma^n(\varphi(p_{11}))\right)$ such that $\varphi(\omega) \leqslant \Gamma^{-n}(\omega)$ when $\omega_0^n \leqslant \omega \leqslant \omega_0$ while $\varphi(\omega) > \Gamma^{-n}(\omega)$ when $p_{01} \leqslant \omega < \underline{\omega}_0^n$

Next, we prove $\omega_0^n < \overline{\omega}_0^n$ by contradiction. Assume $\underline{\omega}_0^n \geqslant \overline{\omega}_0^n$, we have, considering the increasing monotonicity of $\Gamma^{-n}(\omega)$ in ω

$$\varphi\left(\Gamma\left(\overline{\omega}_0^n\right)\right) = \Gamma^{-n}\left(\overline{\omega}_0^n\right) \leqslant \Gamma^{-n}\left(\underline{\omega}_0^n\right) = \varphi\left(\underline{\omega}_0^n\right) \tag{3.38}$$

Since $\Gamma\left(\overline{\omega}_0^n\right) > \Gamma(\Gamma^n(\varphi(p_{01}))) > \Gamma^n(\varphi(p_{11})) > \underline{\omega}_0^n$, we have $\varphi\left(\Gamma\left(\overline{\omega}_0^n\right)\right) > \varphi\left(\underline{\omega}_0^n\right)$ according to the monotonicity of $\varphi(\omega)$, which contradicts (3.38). Hence, we have $\underline{\omega}_0^n < \overline{\omega}_0^n$.■

Based on the two fixed points $\underline{\omega}_0^n$ and $\overline{\omega}_0^n$ in Lemma 3.8, we further divide the region $[\Gamma^n(\varphi(p_{01})), \Gamma^{n+1}(\varphi(p_{01})))$ into the following four subregions:

$$\begin{aligned}\left[\Gamma^n(\varphi(p_{01})), \Gamma^{n+1}(\varphi(p_{01}))\right) = & \left[\Gamma^n(\varphi(p_{01})), \underline{\omega}_0^n\right) \\ & \cup \left[\underline{\omega}_0^n, \bar{\omega}_0^n\right) \\ & \cup \left[\bar{\omega}_0^n, \Gamma^n(\varphi(p_{11}))\right) \\ & \cup \left[\Gamma^n(\varphi(p_{11})), \Gamma^{n+1}(\varphi(p_{01}))\right)\end{aligned}.$$

3.10.1 Region n − *1*

The following proposition quantifies how many time slots are required for a belief value to recover to the given threshold value ω^*.

Proposition 3.3 *When* $\Gamma^n(\varphi(p_{11})) \leqslant \omega^* < \Gamma^{n+1}(\varphi(p_{01}))$, *we have* $L(\Gamma(\varphi(\omega)), \omega^*) = L(\varphi(p_{11}), \omega^*) - 1 = n$ for $\omega \in [p_{01}, p_{11}]$

Proof Since $\Gamma^n(\varphi(p_{11})) \leqslant \omega^*$, we have

$$L(\varphi(p_{11}), \omega^*) \geqslant L(\varphi(p_{11}), \Gamma^n(\varphi(p_{11}))) = n + 1 \tag{3.39}$$

On the other hand, considering $\omega^* < \Gamma^{n+1}(\varphi(p_{11}))$, we have

$$L(\varphi(p_{11}), \omega^*) < L\left(\varphi(p_{11}), \Gamma^{n+1}(\varphi(p_{11}))\right) = n + 2 \tag{3.40}$$

Combining (3.39) and (3.40), we have $L(\varphi(p_{11}), \omega^*) = n + 1$.

Since $\Gamma(\varphi(\omega)) \geqslant \Gamma(\varphi(p_{01}))$, then

$$L(\Gamma(\varphi(\omega)), \omega^*) \leqslant L(\Gamma(\varphi(p_{01})), \omega^*) = n$$

Further, we have $L(\Gamma(\varphi(\omega)), \omega^*) = n$. Hence, the lemma holds. □

By Proposition 3.3, we have the following lemma to linearize $V_{\beta,v}(\omega; \omega, 1)$.

Lemma 3.9 *If* $\Gamma^n(\varphi(p_{11})) \leqslant \omega^* < \Gamma^{n+1}(\varphi(p_{01}))$ *, then we have for* $\omega \in [p_{01}, p_{11}]$

$$\begin{aligned} V_{\beta,v}(\omega; \omega, 1) = (1-\varepsilon)\omega + \beta\big[(1-\varepsilon)\omega V_{\beta,v}(p_{11}; p_{11}) \\ +\varepsilon\omega V_{\beta,v}(p_{11}; \Psi(\omega)) + (1-\omega)V_{\beta,v}(p_{01}; \Psi(\omega))\big], \end{aligned} \tag{3.41}$$

where $L(\Psi(\omega), \omega^*) = n$

$$V_{\beta,v}(p_{11}; p_{11}) \overset{(e1)}{=} V_{\beta,v}(p_{11}; \omega, 1) \tag{3.42}$$

$$V_{\beta,v}(p_{01}; \Psi(\omega)) \overset{(e2)}{=} \frac{1-\beta^n}{1-\beta} v + \beta^n V_{\beta,v}(\Gamma^n(p_{01}); \omega, 1) \tag{3.43}$$

$$V_{\beta,v}(p_{11}; \Psi(\omega)) \overset{(e3)}{=} \frac{1-\beta^n}{1-\beta} v + \beta^n V_{\beta,v}(\Gamma^n(p_{11}); \omega, 1) \tag{3.44}$$

$$\begin{aligned} & V_{\beta,v}(p_{11}; \omega, 1) \\ & \overset{(e4)}{=} (1-\varepsilon)p_{11} + \beta(1-\varepsilon)p_{11} V_{\beta,v}(p_{11}; p_{11}) \\ & +\beta(1-p_{11})V_{\beta,v}(p_{01}; \Psi(\omega)) + \varepsilon\beta p_{11} V_{\beta,v}(p_{11}; \Psi(\omega)) \end{aligned} \tag{3.45}$$

$$\begin{aligned} & V_{\beta,v}(\Gamma^n(p_{01}); \omega, 1) \\ & \overset{(e5)}{=} (1-\varepsilon)\Gamma^n(p_{01}) + \beta(1-\varepsilon)\Gamma^n(p_{01})V_{\beta,v}(p_{11}; p_{11}) \\ & +\beta(1-\Gamma^n(p_{01}))V_{\beta,v}(p_{01}; \Psi(\omega)) + \beta\varepsilon\Gamma^n(p_{01})V_{\beta,v}(p_{11}; \Psi(\omega)) \end{aligned} \tag{3.46}$$

$$\begin{aligned} & V_{\beta,v}(\Gamma^n(p_{11}); \omega, 1) \\ & \overset{(e6)}{=} (1-\varepsilon)\Gamma^n(p_{11}) + \beta(1-\varepsilon)\Gamma^n(p_{11})V_{\beta,v}(p_{11}; p_{11}) \\ & +\beta(1-\Gamma^n(p_{11}))V_{\beta,v}(p_{01}; \Psi(\omega)) + \beta\varepsilon\Gamma^n(p_{11})V_{\beta,v}(p_{11}; \Psi(\omega)) \end{aligned} \tag{3.47}$$

Proof (e1) is due to $L(\Psi(p_{11}), \omega^*) = L(\Psi(\omega), \omega^*)$ by Proposition 3.3. (e2) and (e3) are due to $L(\Psi(\Gamma^n(\Psi(\omega))), \omega^*) = L(\Psi(\omega), \omega^*)$by Proposition 3.3. (e4)–(e6) are due to Lemma 3.6.

Remark 3.5 By (3.42)–(3.47), we can obtain $V_{\beta,v}(p_{11}; p_{11}), V_{\beta,v}(p_{01}; \Psi(\omega))$, and $V_{\beta,\ v}(p_{11}; \Psi(\omega))$ which are substituted in (3.41), leading to the linearization of $V_{\beta,v}(\omega; \omega, 1)$. □

3.10.2 *Region* n − 2

Proposition 3.4 *If* $\overline{\omega}_0^n \leqslant \omega^* < \Gamma^n(\varphi(p_{11}))$, *then*

$$L(\Gamma(\varphi(\omega)), \omega^*) = \begin{cases} L(\varphi(p_{11}), \omega^*) = n, & \text{if } \omega \in [p_{01}, \Gamma(\omega^*)] \\ L(\varphi(p_{11}), \omega^*) - 1 = n - 1, & \text{if } \omega = p_{11} \end{cases}$$

Proof According to Lemma 3.8, we have $\varphi(\Gamma(\omega^*)) \leqslant \Gamma^{-n}(\omega^*)$. Now we prove the lemma by two cases.

(i) when $\omega \in [p_{01}, \Gamma(\omega^*)]$, we have

$$\varphi(\omega) \leqslant \varphi(\Gamma(\omega^*)) \leqslant \Gamma^{-n}(\omega^*)$$

and furthermore,

$$L(\Gamma(\varphi(\omega)), \omega^*) \geqslant L\big(\Gamma^{-n+1}(\omega^*), \omega^*\big) = n$$

Considering $L(\varphi(p_{11}), \omega^*) = n$, we have

$$L(\Gamma(\varphi(\omega)), \omega^*) = L(\varphi(p_{11}), \omega^*) = n$$

(ii) when $\omega = p_{11}$, we have

$$L(\Gamma(\varphi(p_{11})), \omega^*) = L(\varphi(p_{11}), \omega^*) - 1 = n - 1$$

Combining two cases, we obtain the lemma. □

Lemma 3.10 *If* $\overline{\omega}_0^n \leqslant \omega^* < \Gamma^n(\varphi(p_{11}))$, *then we have for* $\omega \in [p_{01}, \Gamma(\omega^*)]$

$$\begin{aligned} V_{\beta,v}(p_{11}; p_{11}) = (1-\varepsilon)p_{11} + \beta\big[(1-\varepsilon)p_{11}V_{\beta,v}(p_{11}; p_{11}) \\ + \varepsilon p_{11}V_{\beta,v}(p_{11}; \Psi(p_{11})) + (1-p_{11})V_{\beta,v}(p_{01}; \Psi(p_{11}))\big] \end{aligned} \tag{3.48}$$

$$\begin{aligned} V_{\beta,v}(\omega; \omega, 1) = (1-\varepsilon)\omega + \beta\big[(1-\varepsilon)\omega V_{\beta,v}(p_{11}; p_{11}) \\ + \varepsilon\omega V_{\beta,v}(p_{11}; \Psi(\omega)) + (1-\omega)V_{\beta,v}(p_{01}; \Psi(\omega))\big] \end{aligned} \tag{3.49}$$

where $L(\varphi(p_{11}), \omega^*) = n$

$$V_{\beta,v}(p_{01}; \Psi(\omega)) \overset{(e1)}{=} \frac{1-\beta^n}{1-\beta}v + \beta^n V_{\beta,v}(\Gamma^n(p_{01}); \omega^*, 1) \tag{3.50}$$

$$V_{\beta,v}(p_{11}; \Psi(\omega)) \overset{(e2)}{=} \frac{1-\beta^n}{1-\beta}v + \beta^n V_{\beta,v}(\Gamma^n(p_{11}); \omega^*, 1) \tag{3.51}$$

$$V_{\beta,v}(p_{01};\Psi(p_{11})) \overset{(e3)}{=} \frac{1-\beta^{n-1}}{1-\beta}v + \beta^{n-1}V_{\beta,v}\left(\Gamma^{n-1}(p_{01});\omega^*,1\right) \tag{3.52}$$

$$V_{\beta,v}(p_{11};\Psi(p_{11})) \overset{(e4)}{=} \frac{1-\beta^{n-1}}{1-\beta}v + \beta^{n-1}V_{\beta,v}\left(\Gamma^{n-1}(p_{11});\omega^*,1\right) \tag{3.53}$$

$$\begin{aligned} &V_{\beta,v}(\Gamma^n(p_{01});\omega^*,1) \\ &\overset{(e5)}{=} (1-\varepsilon)\Gamma^n(p_{01}) + \beta(1-\varepsilon)\Gamma^n(p_{01})V_{\beta,v}(p_{11};p_{11}) \\ &+\beta(1-\Gamma^n(p_{01}))V_{\beta,v}(p_{01};\Psi(\omega^*)) + \beta\varepsilon\Gamma^n(p_{01})V_{\beta,v}(p_{11};\Psi(\omega^*)) \end{aligned} \tag{3.54}$$

$$\begin{aligned} &V_{\beta,v}(\Gamma^n(p_{11});\omega^*,1) \\ &\overset{(e6)}{=} (1-\varepsilon)\Gamma^n(p_{11}) + \beta(1-\varepsilon)\Gamma^n(p_{11})V_{\beta,v}(p_{11};p_{11}) \\ &+\beta(1-\Gamma^n(p_{11}))V_{\beta,v}(p_{01};\Psi(\omega^*)) + \beta\varepsilon\Gamma^n(p_{11})V_{\beta,v}(p_{11};\Psi(\omega^*)) \end{aligned} \tag{3.55}$$

$$\begin{aligned} &V_{\beta,v}\left(\Gamma^{n-1}(p_{01});\omega^*,1\right) \\ &\overset{(e7)}{=} (1-\varepsilon)\Gamma^{n-1}(p_{01}) + \beta(1-\varepsilon)\Gamma^{n-1}(p_{01})V_{\beta,v}(p_{11};p_{11}) \\ &+\beta\left(1-\Gamma^{n-1}(p_{01})\right)V_{\beta,v}(p_{01};\Psi(\omega^*)) + \beta\varepsilon\Gamma^{n-1}(p_{01})V_{\beta,v}(p_{11};\Psi(\omega^*)) \end{aligned} \tag{3.56}$$

$$\begin{aligned} &V_{\beta,v}\left(\Gamma^{n-1}(p_{11});\omega^*,1\right) \\ &\overset{(e8)}{=} (1-\varepsilon)\Gamma^{n-1}(p_{11}) + \beta(1-\varepsilon)\Gamma^{n-1}(p_{11})V_{\beta,v}(p_{11};p_{11}) \\ &+\beta\left(1-\Gamma^{n-1}(p_{11})\right)V_{\beta,v}(p_{01};\Psi(\omega^*)) + \beta\varepsilon\Gamma^{n-1}(p_{11})V_{\beta,v}(p_{11};\Psi(\omega^*)) \end{aligned} \tag{3.57}$$

Proof (e1) and (e2) are due to $L(\Psi(\omega),\omega^*) = L(\Psi'(\omega^*),\omega^*)$ for any $\omega \in [p_{01},\Gamma(\omega^*)] = n$ by Proposition 3.4. (e3) and (e4) are due to $L(\Psi(p_{11}),\omega^*) = L(\Psi(\omega^*),\omega^*) - 1 = n-1$ by Proposition 3.4. (e5)–(e8) are due to Lemma 3.6. □

Remark 3.6 By (3.48), (3.52), (3.53), (3.56), and (3.57), we can obtain $V_{\beta,v}(p_{11};p_{11})$,$V_{\beta,v}(p_{01};\Psi(\omega^*))$ and $V_{\beta,v}(p_{11};\Psi(\omega^*))$. Then plugging them into (3.54), (3.55), (3.50), and (3.51), we have $V_{B,v}(p_{01};\Psi(\omega))$ and $V_{\beta,v}(p_{11};\Psi(\omega))$. Further, the linearization version of $V_{\beta,v}(\omega;\omega,1)$ is obtained.

3.10.3 Region n − *4*

Proposition 3.5 *If* $\Gamma^n(\varphi(p_{01}))\leqslant\omega^* < \underline{\omega}_0^n$ *we have*

$$L(\Gamma(\varphi(\omega)),\omega^*) = L(\varphi(p_{11}),\omega^*) - 1 = n-1$$

for $\omega \in [\omega^*, p_{11}]$

Proof According to the monotonicity of $\varphi(\omega)$, we have $\varphi(\omega) \geqslant \varphi(\omega^*)$ for $\omega \in [\omega^*, p_{11}]$ and $\varphi(\omega^*) > \Gamma^{-n}(\omega^*)$ by Lemma 3.8. Thus, $\varphi(\omega) > \Gamma^{-n}(\omega^*)$ and $L(\Gamma(\varphi(\omega)), \omega^*) < L(\Gamma^{1-n}(\omega^*), \omega^*) = n$. Considering $L(\varphi(p_{11}), \omega^*) = n$, we have $L(\Gamma(\varphi(\omega)), \omega^*) = n - 1$. Therefore, Proposition 3.5 holds. □

Lemma 3.11 *If* $\Gamma^n(\varphi(p_{01})) \leqslant \omega^* < \underline{\omega}_0^n$ *we have for* $\omega \in [\omega^*, p_{11}]$

$$\begin{aligned} V_{\beta,v}(\omega;\omega,1) = (1-\varepsilon)\omega + \beta\big[(1-\varepsilon)\omega V_{\beta,v}(p_{11};p_{11}) \\ +\varepsilon\omega V_{\beta,v}(p_{11};\Psi(\omega)) + (1-\omega)V_{\beta,v}(p_{01};\Psi(\omega))\big] \end{aligned} \tag{3.58}$$

where

$$V_{\beta,v}(p_{11};p_{11}) \overset{(e1)}{=} V_{\beta,v}(p_{11};\omega,1) \tag{3.59}$$

$$V_{\beta,v}(p_{01};\Psi(\omega)) \overset{(e2)}{=} \frac{1-\beta^{n-1}}{1-\beta}v + \beta^{n-1}V_{\beta,v}\big(\Gamma^{n-1}(p_{01});\omega,1\big) \tag{3.60}$$

$$V_{\beta,v}(p_{11};\Psi(\omega)) \overset{(e3)}{=} \frac{1-\beta^{n-1}}{1-\beta}v + \beta^{n-1}V_{\beta,v}\big(\Gamma^{n-1}(p_{11});\omega,1\big) \tag{3.61}$$

$$\begin{aligned} &V_{\beta,v}(p_{11};\omega,1) \\ &\overset{(e4)}{=} (1-\varepsilon)p_{11} + \beta(1-\varepsilon)p_{11}V_{\beta,v}(p_{11};p_{11}) \\ &+\beta(1-p_{11})V_{\beta,v}(p_{01};\Psi(\omega)) + \varepsilon\beta p_{11}V_{\beta,v}(p_{11};\Psi(\omega)) \end{aligned} \tag{3.62}$$

$$\begin{aligned} &V_{\beta,v}\big(\Gamma^{n-1}(p_{01});\omega,1\big) \\ &\overset{(e5)}{=} (1-\varepsilon)\Gamma^{n-1}(p_{01}) + \beta(1-\varepsilon)\Gamma^{n-1}(p_{01})V_{\beta,v}(p_{11};p_{11}) \\ &+\beta\big(1-\Gamma^{n-1}(p_{01})\big)V_{\beta,v}(p_{01};\Psi(\omega)) + \beta\varepsilon\Gamma^{n-1}(p_{01})V_{\beta,v}(p_{11};\Psi(\omega)) \end{aligned} \tag{3.63}$$

$$\begin{aligned} &V_{\beta,v}\big(\Gamma^{n-1}(p_{11});\omega,1\big) \\ &\overset{(e6)}{=} (1-\varepsilon)\Gamma^{n-1}(p_{11}) + \beta(1-\varepsilon)\Gamma^{n-1}(p_{11})V_{\beta,v}(p_{11};p_{11}) \\ &+\beta\big(1-\Gamma^{n-1}(p_{11})\big)V_{\beta,v}(p_{01};\Psi(\omega)) + \beta\varepsilon\Gamma^{n-1}(p_{11})V_{\beta,v}(p_{11};\Psi(\omega)) \end{aligned} \tag{3.64}$$

Proof (e1)–(e3) are due to $L(\Psi(\omega), \omega^*) = n - 1$ for any $\omega \in [\omega^*, p_{11}]$ by Proposition 3.5. (e4)–(e6) are due to Lemma 3.6. □

Remark 3.7 By (3.59)–(3.64), we can obtain $V_{\beta,v}(p_{11};p_{11}), V_{\beta,v}(p_{01};\Psi(\omega))$ and $V_{\beta,v}(p_{11};\Psi(\omega))$. Plugging them into (3.58), we get the linearized version $V_{\beta,v}(\omega;\omega,1)$.

3.10.4 *Region* n-3

For the region $[\underline{\omega}_0^n, \overline{\omega}_0^n)$, by theoretic analysis, we find out that there exist an infinite number of fixed points so that this region can be further divided into an infinite number of subregions. For each subregion, we can compute the subsidy v by the above similar way.

Considering computation complexity, we simply use linear interpolation to compute the index in this region for two practical reasons:

- the Whittle approach is an approximate one in essence,
- the nonlinearity shows the tradeoff between precise computation and computation cost.

3.10.5 *Region 5*

For $\omega_0 < \omega^* < p_{11}$, it holds that $L(p_{11}, \omega^*) = 0$ and $L(\Gamma(\varphi(\omega)), \omega^*) = \infty$ for $\omega \in [p_{01}, p_{11}]$. Therefore, we have $V_{\beta,v}(\Gamma(\varphi(\omega)); \Gamma(\varphi(\omega))) = \frac{v}{1-\beta}$.

3.11 Index Computation for Negatively Correlated Channels

Since the nonlinear part $V_{\beta,v}(\omega^*; \omega^*, 1)$ for different regions has been linearized in previous sections, we now begin to compute the Whittle index based on the balance equation as follows:

$$V_{\beta,v}(\omega^*; \omega^*, 0) = V_{\beta,v}(\omega^*; \omega^*, 1) \tag{3.65}$$

3.11.1 *Region 1*

When $p_{11} \leqslant \omega^* < \omega_0$, we know that $L(\Gamma(\varphi(\omega)), \omega^*) = 0$ for any $\omega \in [p_{11}, p_{01}]$ according to

Proposition 3.2 Thus, we have

$$\begin{aligned} & V_{\beta,v}(\omega^*; \omega^*, 0) \\ = & \, v + \beta V_{\beta,v}(\Gamma(\omega^*); \Gamma(\omega^*)) \end{aligned}$$

$$
\begin{aligned}
&= v + \beta(1-\varepsilon)\Gamma(\omega^*) + \beta^2\big[(1-\varepsilon)\Gamma(\omega^*)V_{\beta,v}(p_{11};p_{11}) \\
&\quad +(1-(1-\varepsilon)\Gamma(\omega^*))V_{\beta,v}(\Psi(\Gamma(\omega^*));\Psi(\Gamma(\omega^*)))\big] \\
&= v + \beta(1-\varepsilon)\Gamma(\omega^*) + \beta^2\big[(1-\varepsilon)\Gamma(\omega^*)V_{\beta,v}(p_{11};p_{11}]) \\
&\quad +\varepsilon\Gamma(\omega^*)V_{\beta,v}(p_{11};\Psi(\Gamma(\omega^*))) + (1-\Gamma(\omega^*))V_{\beta,v}(p_{01};\Psi(\Gamma(\omega^*)))\big] \\
&= v + \beta(1-\varepsilon)\Gamma(\omega^*) + \beta^2\big[(1-\varepsilon)\Gamma(\omega^*)V_{\beta,v}(p_{11};p_{11}) \\
&\quad +\varepsilon\Gamma(\omega^*)V_{\beta,v}(p_{11};\Psi(\omega^*)) + (1-\Gamma(\omega^*))V_{\beta,v}(p_{01};\Psi(\omega^*))\big]
\end{aligned} \tag{3.66}
$$

According to Lemma 3.7, we have

$$
\begin{aligned}
V_{\beta,v}(\omega^*;\omega^*,1) &= (1-\varepsilon)\omega^* + \beta\big[(1-\varepsilon)\omega^* V_{\beta,v}(p_{11};p_{11}) \\
&\quad +\varepsilon\omega^* V_{\beta,v}(p_{11};\Psi(\omega^*)) + (1-\omega^*)V_{\beta,v}(p_{01};\Psi(\omega^*))\big]
\end{aligned} \tag{3.67}
$$

According to (3.65), combined with (3.35), (3.36), and (3.37) by letting $\omega = \omega^*$, we have, in the matrix form, the following linear functions:

$$
\widetilde{\mathbf{M}}_1 \cdot \begin{pmatrix} V_{\beta,v}(p_{11};p_{11}) \\ V_{\beta,v}(p_{11};\Psi(\omega^*)) \\ V_{\beta,v}(p_{01};\Psi(\omega^*)) \\ v \end{pmatrix} = \begin{pmatrix} \beta(\varepsilon-1)\Gamma(p_{11}) \\ (\varepsilon-1)p_{11} \\ (\varepsilon-1)p_{01} \\ (\varepsilon-1)(\omega^*-\beta\Gamma(\omega^*)) \end{pmatrix} \tag{3.68}
$$

where $\widetilde{\mathbf{M}}_1$ is defined in (3.69).

$$
\widetilde{\mathbf{M}}_1 = \begin{pmatrix} \beta^2(1-\epsilon)\Gamma(p_{11})-1 & \beta^2\epsilon\Gamma(p_{11}) & \beta^2(1-\Gamma(p_{11})) & 1 \\ \beta(1-\epsilon)p_{11} & \beta\epsilon p_{11}-1 & \beta(1-p_{11}) & 0 \\ \beta(1-\epsilon)p_{01} & \beta\epsilon p_{01} & \beta(1-p_{01})-1 & 0 \\ \beta(1-\epsilon)(\omega^*-\beta\Gamma(\omega^*)) & \beta\epsilon(\omega^*-\beta\Gamma(\omega^*)) & \beta[(1-\beta)-(\omega^*-\beta\Gamma(\omega^*))] & -1 \end{pmatrix} \tag{3.69}
$$

Thus, by some mathematical operations, we obtain the Whittle index v for the region $p_{11} \leqslant \omega^* < \omega_0$

$$
v = \frac{\beta(1-\varepsilon)p_{01} + (1-\varepsilon)(\omega^*-\beta\Gamma(\omega^*))}{1+\beta(p_{01}-\varepsilon p_{11}) - \beta(1-\varepsilon)(\beta\Gamma(p_{11})+\omega^*-\beta\Gamma(\omega^*))} \tag{3.70}
$$

3.11.2 Region 2

When $\omega_0 < \omega^* < \Gamma(p_{11})$, we know that $L(\Gamma(\varphi(\omega)), \omega^*) = 0$ for any $\omega \in [p_{11}, p_{01}]$.

Thus, we have

$$\begin{aligned}&V_{\beta,\nu}(\omega^*;\omega^*,0)\\&=\nu+\beta V_{\beta,\nu}(\Gamma(\omega^*);\Gamma(\omega^*))\\&=\nu+\beta\nu+\beta^2V_{\beta,\nu}(\Gamma^2(\omega^*);\Gamma^2(\omega^*))\\&=\frac{\nu}{1-\beta}\end{aligned}$$

Meanwhile, by Corollary 3.4, we have $V_{\beta,\nu}(\omega^*;\omega^*,1)$. Further, combined with the balance equation at co', we have analogously the following matrix function:

$$\widetilde{\mathbf{M}}_2\cdot\begin{pmatrix}V_{\beta,\nu}(p_{11};p_{11})\\V_{\beta,\nu}(p_{11};\Psi(\omega^*))\\V_{\beta,\nu}(p_{01};\Psi(\omega^*))\\\nu\end{pmatrix}=\begin{pmatrix}-\beta(1-\varepsilon)\Gamma(p_{11})\\-(1-\varepsilon)p_{11}\\-(1-\varepsilon)p_{01}\\-(1-\varepsilon)\omega^*\end{pmatrix}\tag{3.71}$$

where $\widetilde{\mathbf{M}}_2$ is defined in (3.72).

$$\widetilde{\mathbf{M}}_1=\begin{pmatrix}\beta^2(1-\epsilon)\Gamma(p_{11})-1&\beta^2\epsilon\Gamma(p_{11})&\beta^2(1-\Gamma(p_{11}))&1\\\beta(1-\epsilon)p_{11}&\beta\epsilon p_{11}-1&\beta(1-p_{11})&0\\\beta(1-\epsilon)p_{01}&\beta\epsilon p_{01}&\beta(1-p_{01})-1&0\\\beta(1-\epsilon)\omega^*&\beta\epsilon\omega^*&\beta(1-\omega^*)&-\frac{1}{1-\beta}\end{pmatrix}\tag{3.72}$$

Thus, by solving the matrix function, we obtain the following Whittle index for the region $\omega_0<\omega^*<\Gamma(p_{11})$:

$$\nu=\frac{\beta(1-\varepsilon)p_{01}+(1-\varepsilon)(1-\beta)\omega^*}{1+\beta(p_{01}-\varepsilon p_{11})-\beta(1-\varepsilon)(\beta\Gamma(p_{11})+(1-\beta)\omega^*)}\tag{3.73}$$

3.11.3 Region 3

For $\Gamma(p_{11})\leqslant\omega^*<\overline{\omega}_0$, we compute ν in the following.

Proposition 3.6 *When $\Gamma(p_{11})\leqslant\omega^*<p_{01}$, $L(\Gamma(\omega),\omega^*)=\infty$ for any $\omega\in[p_{11},p_{01}]$.*

Proof Based on the decreasing monotonicity of $\Gamma(\omega)$, then $\Gamma(\omega)\leqslant\Gamma(p_{11})\leqslant\omega^*$ for any $\omega\in[p_{11},p_{01}]$. Considering $\Gamma(\omega)\in[p_{11},p_{01}]$, then $\Gamma^2(\omega)\leqslant\Gamma(p_{11})\leqslant\omega^*$, and so on; that is, $L(\Gamma(\omega),\omega^*)=\infty$

When $\Gamma(p_{11})\leqslant\omega^*<\overline{\omega}_0$, by some mathematical operations, we can obtain the following matrix function:

$$\widetilde{\mathbf{M}}_3 \cdot \begin{pmatrix} V_{\beta,v}(p_{11};p_{11}) \\ V_{\beta,v}(p_{11};\Psi(\omega^*)) \\ V_{\beta,v}(p_{01};\Psi(\omega^*)) \\ v \end{pmatrix} = \begin{pmatrix} 0 \\ -(1-\varepsilon)p_{11} \\ -(1-\varepsilon)p_{01} \\ -(1-\varepsilon)\omega^* \end{pmatrix} \tag{3.74}$$

where $\widetilde{\mathbf{M}}_3$ is defined in (3.75).

$$\widetilde{\mathbf{M}}_1 = \begin{pmatrix} -1 & 0 & 0 & \frac{1}{1-\beta} \\ \beta(1-\epsilon)p_{11} & \beta\epsilon p_{11}-1 & \beta(1-p_{11}) & 0 \\ \beta(1-\epsilon)p_{01} & \beta\epsilon p_{01} & \beta(1-p_{01})-1 & 0 \\ \beta(1-\epsilon)\omega^* & \beta\epsilon\omega^* & \beta(1-\omega^*) & -\frac{1}{1-\beta} \end{pmatrix} \tag{3.75}$$

Therefore, we have the Whittle index v for the region $\Gamma(p_{11}) \leqslant \omega^* < \omega_0$ as follows:

$$v = \frac{\beta(1-\varepsilon)p_{01} + (1-\beta)(1-\varepsilon)\omega^*}{1+\beta(p_{01}-\varepsilon p_{11}) - \beta(1-\varepsilon)\omega^*} \tag{3.76}$$

3.11.4 Region 4

When $\overline{\omega}_0 \leqslant \omega^* < p_{01}$, we have $L(\Psi(\omega), \omega^*) = \infty$ and $L(p_{11}, \omega^*) = \infty$ by Corollary 3.3. Thus, $V_{\beta,v}(p_{11};p_{11}) = \frac{v}{1-\beta}$ and $V_{\beta,v}(\Psi(\omega^*);\Psi(\omega^*)) = \frac{v}{1-\beta}$

$$\begin{aligned} & V_{\beta,v}(\omega^*;\omega^*,0) \\ & = v + \beta V_{\beta,v}(\Gamma(\omega^*);\Gamma(\omega^*)) \\ & = v + \beta m + \beta^2 V_{\beta,v}(\Gamma^2(\omega^*);\Gamma^2(\omega^*)) \\ & = \frac{v}{1-\beta} \end{aligned}$$

$$\begin{aligned} & V_{\beta,v}(\omega^*;\omega^*,1) \\ & = (1-\varepsilon)\omega^* + \beta[(1-\varepsilon)\omega^* V_{\beta,v}(p_{11};p_{11}) \\ & +(1-(1-\varepsilon)\omega^*)V_{\beta,v}(\Psi(\omega^*);\Psi(\omega^*))] \\ & = (1-\varepsilon)\omega^* + \beta\left[(1-\varepsilon)\omega^* \frac{v}{1-\beta} + (1-(1-\varepsilon)\omega^*)\frac{v}{1-\beta}\right] \\ & = (1-\varepsilon)\omega^* + \beta\frac{v}{1-\beta} \end{aligned}$$

Therefore, based on the balance equation $V_{\beta,v}(\omega^*;\omega^*,0) = V_{\beta,v}(\omega^*;\omega^*,1)$, we have

$$v = (1-\varepsilon)\omega^* \tag{3.77}$$

Combing v in (3.70), (3.73), (3.76) with (3.77), we finally obtain the Whittle index shown in (3.19) for the negatively correlated channels.

3.12 Index Computation for Positively Correlated Channels

3.12.1 Region 1

When $\Gamma^n(\varphi(p_{11})) \leqslant \omega_\beta^*(m) < \Gamma^{n+1}(\varphi(p_{01}))$, according to Lemma 3.9, we let $\omega = \omega^*$ for (3.41)–(3.47), combine the balance function $V_{\beta,v}(\omega^*;\omega^*,0) = V_{\beta,v}(\omega^*;\omega^*,1)$ at ω^* and the following equation:

$$\begin{aligned}
&V_{\beta,v}(\omega^*;\omega^*,0)\\
&= v + \beta V_{\beta,v}(\Gamma(\omega^*);\Gamma(\omega^*))\\
&= v + \beta V_{\beta,v}(\Gamma(\omega^*);\omega^*,1)\\
&= v + \beta(1-\varepsilon)\Gamma(\omega^*) + \beta\big[(1-\varepsilon)\Gamma(\omega^*)V_{\beta,v}(p_{11};p_{11})\\
&\quad + (1-\Gamma(\omega^*))V_{\beta,v}(p_{01};\Psi(\omega^*)) + \varepsilon\Gamma(\omega^*)V_{\beta,v}(p_{11};\Psi(\omega^*))\big]
\end{aligned} \tag{3.78}$$

Finally, we can obtain the following linear functions:

$$\mathbf{M}_1 \cdot \begin{pmatrix} V_{\beta,v}(p_{11};p_{11}) \\ V_{\beta,v}(p_{11};\Psi(\omega^*)) \\ V_{\beta,v}(p_{01};\Psi(\omega^*)) \\ v \end{pmatrix} = \begin{pmatrix} -(1-\varepsilon)p_{11} \\ -\beta^n(1-\varepsilon)\Gamma^n(p_{11}) \\ -\beta^n(1-\varepsilon)\Gamma^n(p_{01}) \\ (1-\varepsilon)(\omega^* - \beta\Gamma(\omega^*)) \end{pmatrix} \tag{3.79}$$

where $\mathbf{M}_1$ is defined in (3.80)

$$\mathbf{M}_1 = \begin{pmatrix} -1 & 0 & 0 & \frac{1}{1-\beta} \\ \beta(1-\epsilon)p_{11} & \beta\epsilon p_{11} - 1 & \beta(1-p_{11}) & 0 \\ \beta(1-\epsilon)p_{01} & \beta\epsilon p_{01} & \beta(1-p_{01}) - 1 & 0 \\ \beta(1-\epsilon)\omega^* & \beta\epsilon\omega^* & \beta(1-\omega^*) & -\frac{1}{1-\beta} \end{pmatrix} \tag{3.80}$$

Thus, by simply solving (3.79), we have

$$v = \frac{(1-\varepsilon)\big(1-\beta^{n+1}\big)(\omega^* - \beta\Gamma(\omega^*)) + C_6}{(1-\varepsilon)\big(\beta-\beta^{n+1}\big)(\omega^* - \beta\Gamma(\omega^*)) + C_9} \tag{3.81}$$

3.12.2 Region 2

When $\overline{\omega}_0^n \leqslant \omega_\beta^*(m) < \Gamma^n(\varphi(p_{11}))$, letting $\omega = \omega^*$ for (3.48) – (3.57), combined with (3.78) and the balance function at ω^* we have

$$\mathbf{M}_2 \cdot \begin{pmatrix} V_{\beta,v}(p_{11}; p_{11}) \\ V_{\beta,v}(p_{11}; \Psi(\omega^*)) \\ V_{\beta,v}(p_{01}; \Psi(\omega^*)) \\ V_{\beta,v}(p_{11}; \Psi(p_{11})) \\ V_{\beta,v}(p_{01}; \Psi(p_{11})) \\ v \end{pmatrix} = \begin{pmatrix} -(1-\varepsilon)p_{11} \\ -\beta^{n-1}(1-\varepsilon)\Gamma^{n-1}(p_{11}) \\ -\beta^{n-1}(1-\varepsilon)\Gamma^{n-1}(p_{01}) \\ -\beta^{n}(1-\varepsilon)\Gamma^{n}(p_{11}) \\ -\beta^{n}(1-\varepsilon)\Gamma^{n}(p_{01}) \\ (1-\varepsilon)(\omega^* - \beta\Gamma(\omega^*)) \end{pmatrix} \tag{3.82}$$

where $\mathbf{M}_2$ is defined in (3.83).

$$\mathbf{M}_2 = \begin{pmatrix} \beta(1-\varepsilon)p_{11}-1 & \beta\varepsilon p_{11} & \beta(1-p_{11}) & 0 & 0 & 0 \\ \beta^n(1-\varepsilon)\Gamma^{n-1}(p_{11}) & -1 & 0 & \beta^n\varepsilon\Gamma^{n-1}(p_{11}) & \beta^n(1-\Gamma^{n-1}(p_{11})) & \frac{1-\beta^{n-1}}{1-\beta} \\ \beta^n(1-\varepsilon)\Gamma^{n-1}(p_{01}) & 0 & -1 & \beta^n\varepsilon\Gamma^{n-1}(p_{01}) & \beta^n(1-\Gamma^{n-1}(p_{01})) & \frac{1-\beta^{n-1}}{1-\beta} \\ \beta^{n+1}(1-\varepsilon)\Gamma^n(p_{11}) & 0 & 0 & \beta^{n+1}\varepsilon\Gamma^n(p_{11})-1 & \beta^{n+1}(1-\Gamma^n(p_{11})) & \frac{1-\beta^h}{1-\beta} \\ \beta^{n+1}(1-\varepsilon)\Gamma^n(p_{01}) & 0 & 0 & \beta^{n+1}\varepsilon\Gamma^n(p_{01}) & \beta^{n+1}(1-\Gamma^n(p_{01}))-1 & \frac{1-\beta^n}{1-\beta} \\ \beta(1-\varepsilon)(\omega^*-\beta\Gamma(\omega^*)) & 0 & 0 & \beta\varepsilon(\omega^*-\beta\Gamma(\omega^*)) & \beta(1-\omega^*)-\beta^2(1-\Gamma(\omega^*) & -1 \end{pmatrix} \tag{3.83}$$

Thus

$$v = \frac{(1-\varepsilon)\big(1-\beta^{n+1}\big)(\omega^* - \beta\Gamma(\omega^*)) + C_6}{C_0(\omega^* - \beta\Gamma(\omega^*)) + C_7} \tag{3.84}$$

3.12.3 Region 3

When $\Gamma^{n+1}(\varphi(p_{01})) \leqslant \omega_\beta^*(m) < \underline{\omega}_0^{n+1}$, we have

$$\mathbf{M}_1 \cdot \begin{pmatrix} V_{\beta,v}(p_{11};p_{11}) \\ V_{\beta,v}(p_{11};\omega^*]) \\ V_{\beta,v}(p_{01};\omega^*) \\ v \end{pmatrix} = \begin{pmatrix} -(1-\varepsilon)p_{11} \\ -\beta^{L-1}(1-\varepsilon)\Gamma^{L-1}(p_{11}) \\ -\beta^{L-1}(1-\varepsilon)\Gamma^{L-1}(p_{01}) \\ (1-\varepsilon)(\omega^* - \beta\Gamma(\omega^*)) \end{pmatrix} \tag{3.85}$$

where, $\mathbf{M}_1$ is defined in (3.80). Thus

$$v = \frac{(1-\varepsilon)\big(1-\beta^{n+1}\big)(\omega^* - \beta\Gamma(\omega^*)) + C_6}{(1-\varepsilon)\big(\beta-\beta^{n+1}\big)(\omega^* - \beta\Gamma(\omega^*)) + C_9} \tag{3.86}$$

3.12.4 Region 4

When $p_{01} \le \omega^* < \underline{\omega}_0^1$ we have $L(\Psi(\omega^*), \omega^*) = 0$ since $\Psi(\omega^*) > \omega^*$ according to Lemma 3.8 and $L(p_{11}, \omega^*) = L(\Gamma(\omega^*), \omega^*) = 0$ since $p_{11} > \Gamma(\omega^*) > \omega^*$. Thus,

$$\begin{aligned}
V_{\beta,v}(\omega^*;\omega^*,0) &= v + \beta V_{\beta,v}(\Gamma(\omega^*);\Gamma(\omega^*)) \\
&= v + \beta\big[\omega^* V_{\beta,v}(\Gamma(1);p_{11}) + (1-\omega^*)V_{\beta,v}(\Gamma(0);p_{11})\big] \\
&= v + \beta\big[\omega^* V_{\beta,v}(p_{11};p_{11}) + (1-\omega^*)V_{\beta,v}(p_{01};p_{11})\big] \\
V_{\beta,v}(\omega^*;\omega^*,1) &= (1-\varepsilon)\omega^* + \beta\big[(1-\varepsilon)\omega^* V_{\beta,v}(p_{11};p_{11}) \\
&\quad + (1-(1-\varepsilon)\omega^*)V_{\beta,v}(\Psi(\omega^*);\Psi(\omega^*)) \\
&= (1-\varepsilon)\omega^* + \beta\big[(1-\varepsilon)\omega^* V_{\beta,v}(p_{11};p_{11}) \\
&\quad + \varepsilon\omega^* V_{\beta,v}(p_{11};p_{11}) + (1-\omega^*)V_{\beta,v}(p_{01};p_{11})\big] \\
&= (1-\varepsilon)\omega^* + \beta\big[\omega^* V_{\beta,v}(p_{11};p_{11}) + (1-\omega^*)V_{\beta,v}(p_{01};p_{11})\big]
\end{aligned}$$

Further, based on $V_{\beta,v}(\omega^*;\omega^*,0) = V_{\beta,v}(\omega^*;\omega^*,1)$, we have

$$v = (1-\varepsilon)\omega^* \tag{3.87}$$

3.12.5 Region 5

When $\underline{\omega}_0^n \le \omega^* < \overline{\omega}_0^n$, we obtain v by linear interpolation on two endpoints $v\big(\underline{\omega}_0^n\big)$ and $v\big(\overline{\omega}_0^n\big)$

$$v = v\left(\underline{\omega}_0^n\right) + \left(\omega - \underline{\omega}_0^n\right) \frac{v\left(\overline{\omega}_0^n\right) - v\left(\underline{\omega}_0^n\right)}{\overline{\omega}_0^n - \underline{\omega}_0^n} \tag{3.88}$$

3.12.6 Region 6

When $\omega_0 \leq \omega^* \leq p_{11}$, we have $L(\Gamma^i(\omega^*), \omega^*) = \infty$ for $i \geq 1$, $L(\Psi(p_{11}), \omega^*) = \infty$, and $L(\Psi(\omega^*), \omega^*) = \infty$. Thus,

$$\begin{aligned}
V_{\beta,v}(p_{11}; p_{11}) &= (1-\varepsilon)p_{11} + \beta\Big[(1-\varepsilon)p_{11}V_{\beta,v}(p_{11}; p_{11}) \\
&\quad + (1-(1-\varepsilon)p_{11})V_{\beta,v}(\Psi(p_{11}); \Psi(p_{11}))\Big] \\
&= (1-\varepsilon)p_{11} + \beta\Big[(1-\varepsilon)p_{11}V_{\beta,v}(p_{11}; p_{11}) \\
&\quad + (1-(1-\varepsilon)p_{11})V_{\beta,v}(\Psi(p_{11}); \omega^*, 0)\Big] \\
&= (1-\varepsilon)p_{11} + \beta\Big[(1-\varepsilon)p_{11}V_{\beta,v}(p_{11}; p_{11}) \\
&\quad + (1-(1-\varepsilon)p_{11})\frac{v}{1-\beta}\Big]
\end{aligned} \tag{3.89}$$

$$\begin{aligned}
V_{\beta,v}(\omega^*; \omega^*, 0) &= v + \beta V_{\beta,v}(\Gamma(\omega^*); \Gamma(\omega^*)) \\
&= v + \beta V_{B,v}(\Gamma(\omega^*); \omega^*, 0) \\
&= v + \beta\big[v + \beta V_{\beta,v}(\Gamma^2(\omega^*); \omega^*, 0)\big] \\
&= \frac{v}{1-\beta}
\end{aligned} \tag{3.90}$$

$$\begin{aligned}
V_{\beta,v}(\omega^*; \omega^*, 1) &= (1-\varepsilon)\omega^* + \beta\Big[(1-\varepsilon)\omega^* V_{\beta,v}(p_{11}; p_{11}) \\
&\quad + (1-(1-\varepsilon)\omega^*)V_{\beta,v}(\Psi(\omega^*); \Psi(\omega^*))\Big] \\
&= (1-\varepsilon)\omega^* + \beta\Big[(1-\varepsilon)\omega^* V_{\beta,v}(p_{11}; p_{11}) \\
&\quad + (1-(1-\varepsilon)\omega^*)V_{\beta,v}(\Psi(\omega^*); \omega^*, 0)\Big]
\end{aligned}$$

$$= (1-\varepsilon)\omega^* + \beta\Big[(1-\varepsilon)\omega^* V_{\beta,v}(p_{11};p_{11}) \\ +(1-(1-\varepsilon)\omega^*)\frac{v}{1-\beta}\Big] \tag{3.91}$$

Finally, based on $V_{\beta,v}(\omega^*;\omega^*,0) = V_{\beta,v}(\omega^*;\omega^*,1)$ we obtain the following:

$$v = \frac{(1-\varepsilon)\omega^*}{1-\beta(1-\varepsilon)(p_{11}-\omega^*)} \tag{3.92}$$

Combing v in (3.81), (3.84), (3.86), (3.87), (3.88) with (3.92), we finally obtain the Whittle index shown in (3.20) for the positively correlated channels.

3.13 Numerical Study

In this section, we evaluate the performance of the Whittle index policy by comparing with optimal policy and myopic policy.

3.13.1 Whittle Index versus Optimal Policy

In the first scenario, the parameters are set as $N = 3$, $\varepsilon_i = 0.01$, and $\beta = 1$,$\left\{\left(p_{01}^{(i)}, p_{11}^{(i)}\right)\right\}_{i=1}^{3} = \{(0.3, 0.7), (0.4, 0.8), (0.5, 0.7)\}$. From Fig. 3.5, we observe that the Whittle index policy has almost the same performance with the optimal policy.

In the second scenario, the parameters are set as $N = 3$, $\left\{\left(p_{01}^{(i)}, p_{11}^{(i)}\right)\right\}_{i=1}^{3} = \{(0.3, 0.7), (0.8, 0.4), (0.3, 0.6)\}$, $\varepsilon_i = 0.01$, and $\beta = 1$. From Fig. 3.6, we observe that the Whittle index policy has about 1% performance loss compared with the optimal policy. Combining Figs. 3.5 and 3.6, we have the following intuition result: the Whittle index policy performs worse with the increasing heterogeneity among channels.

3.13.2 Whittle Index verse Myopic Policy

In this scenario $N = 10$, $\left\{\left(p_{01}^{(i)}, p_{11}^{(i)}\right)\right\}_{i=1}^{10}$, $= \{(0.3, 0.9), (0.8, 0.1), (0.30.8),$ $(0.1, 0.9), (0.9, 0.1), (0.4, 0.8), (0.5, 0.3), (0.3, 0.3), (0.3, 0.6), (0.8, 0.1)\}$, and

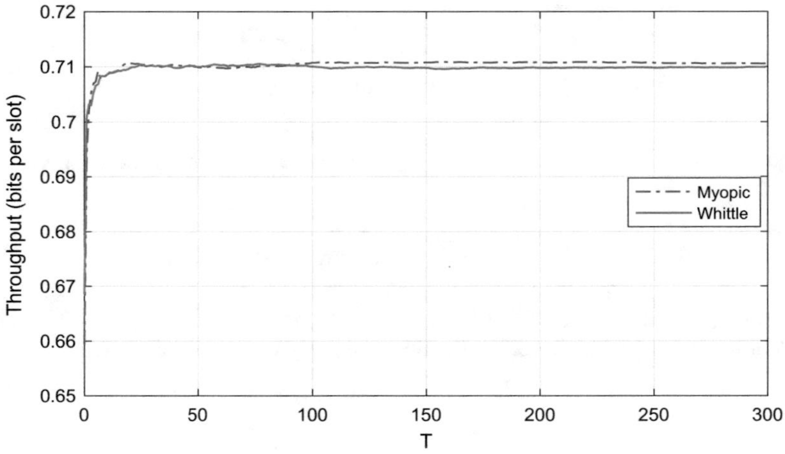

Fig. 3.5 $N = 3, \beta = 1, \varepsilon_i = 0.01, \left\{\left(p_{01}^{(i)}, p_{11}^{(i)}\right)\right\}_{i=1}^{3} = \left\{(0.3, 0.7), (0.4, 0.8), (0.5, 0.7)\right\}$

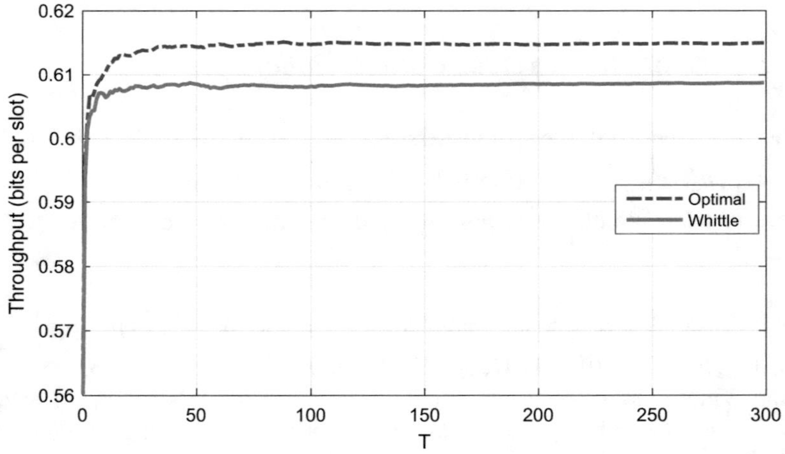

Fig. 3.6 $N = 3, \beta = 1, \varepsilon_i = 0.01, \left\{\left(p_{01}^{(i)}, p_{11}^{(i)}\right)\right\}_{i=1}^{3} = \left\{(0.3, 0.7), (0.8, 0.4), (0.3, 0.6)\right\}$

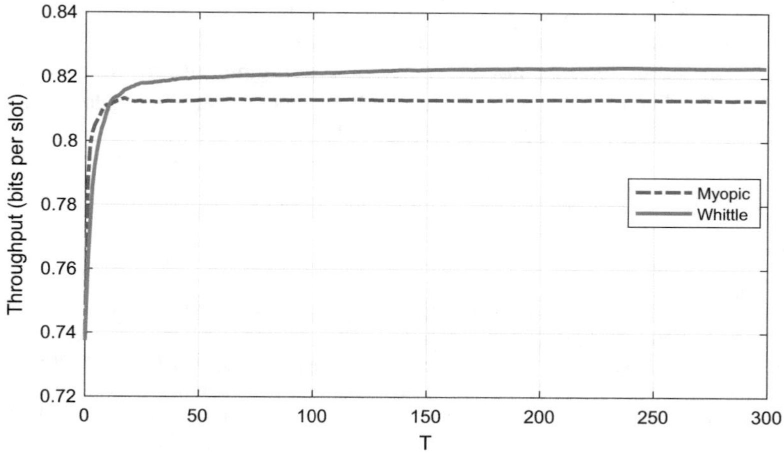

Fig. 3.7 $N = 10, \beta = 1, \varepsilon_i = 0.01, \left\{(p_{01}^{(i)}, p_{11}^{(i)})\right\}_{i=1}^{10} = \left\{(0.3, 0.9), (0.8, 0.1), (0.3, 0.8), (0.1, 0.9),\right.$ $\left.(0.9, 0.1), (0.4, 0.8), (0.5, 0.3), (0.3, 0.3), (0.3, 0.6), (0.8, 0.1)\right\}$

$\varepsilon_i = 0.01$, and $\beta = 1$. From Fig. 3.7, we can see that the Whittle index policy performs a little worse than the myopic policy when $T \leqslant 18$, while after that threshold time, performs better. This can be easily explained as follows: the myopic policy performs better in the initial period since it only exploits information to maximize utility but ignores exploring information for future decision. However, the Whittle index considers the balance between exploitation and exploration, so it performs better after the initial period.

3.14 Summary

In this chapter, we address the multichannel opportunistic access problem, in which a user decides, at each time slot, which channel to access among multiple Gilbert-Elliot channels in order to maximize his aggregated utility, given that the observation of channel state is error-prone. The problem can be cast into a restless multiarmed bandit problem which is proved to be PSPACE-Hard. We thus study the alternative approach, Whittle index policy, a very popular heuristic for restless bandits, which is provably optimal asymptotically and has good empirical performance. However, in the case of imperfect observation, the traditional approach of computing the Whittle index policy cannot be applied because the channel state belief evolution is no more linear, thus rendering the indexability of our problem open. To bridge the gap, we mathematically establish the indexability and establish the closed-form Whittle index, based on which index policy can be constructed. The major technique is our analysis is a fixed point-based approach which enable us to divide the belief information space into a series of regions and then establish a set of periodic structures of the underlying nonlinear dynamic evolving system, based on which we devise the linearization scheme for each region to establish indexability and compute the Whittle index for each region.

References

1. E.N. Gilbert, Capacity of a burst-noise channel. Bell Syst. Tech. J. **39**(5), 1253–1265 (1960)
2. K. Wang, L. Chen, Q. Liu, K.A. Agha, On optimality of myopic sensing policy with imperfect sensing in multi-channel opportunistic access. IEEE Trans. Commun. **61**(9), 3854–3862 (2013)
3. K. Wang, Q. Liu, F. Li, L. Chen, X. Ma, Myopic policy for opportunistic access in cognitive radio networks by exploiting primary user feedbacks. IET Commun. **9**(7), 1017–1025 (2015)
4. J. Nino-Mora, S.S. Villar. Sensor scheduling for hunting elusive hiding targets via whittle's restless bandit index policy. in *Proc. of NetGCoop*, Paris, France, Oct 2011, pp. 1–8.
5. C.H. Papadimitriou, J.N. Tsitsiklis, The complexity of optimal queueing network control. Math. Oper. Res. **24**(2), 293–305 (1999)
6. S. Ahmad, M. Liu, T. Javidi, Q. Zhao, B. Krishnamachari, Optimality of myopic sensing in multichannel opportunistic access. IEEE Trans. Inf. Theory **55**(9), 4040–4050 (2009)
7. K. Wang, L. Chen, On optimality of myopic policy for restless multi-armed bandit problem: an axiomatic approach. IEEE Trans. Signal Process. **60**(1), 300–309 (2012)
8. Y. Liu, M. Liu, S.H.A. Ahmad, Sufficient conditions on the optimality of myopic sensing in opportunistic channel access: a unifying framework. IEEE Trans. Inf. Theory **60**(8), 4922–4940 (2014)
9. Y. Ouyang, D. Teneketzis, On the optimality of myopic sensing in multi-state resources. IEEE Trans. Inf. Theory **60**(1), 681–696 (2014)
10. K. Liu, Q. Zhao, B. Krishnamachari, Dynamic multichannel access with imperfect channel state detection. IEEE Trans. Signal Process. **58**(5), 2795–2807 (2010)
11. K. Wang, L. Chen, Q. Liu, Opportunistic spectrum access by exploiting primary user feedbacks in underlay cognitive radio systems: an optimality analysis. IEEE J. Sel. Topics Signal Process. **7**(5), 869–882 (2013)
12. I.M. Verloop, Asymptotically optimal priority policies for indexable and non-indexable restless bandits. Ann. Appl. Probab. **26**(4), 1947–1995 (2016)
13. R.R. Weber, G. Weiss, On an index policy for restless bandits. J. Appl. Probab. **27**(3), 637–648 (1990)
14. P.R. Singh, X. Guo. Index policies for optimal mean-variance trade-off of inter-delivery times in real-time sensor networks. In *Proc. of 2015 IEEE Conference on Computer Communications (INFOCOM)*, 2015.
15. J. L. Ny, M. Dahleh, E. Feron. Multi-UAV dynamic routing with partial observations using restless bandit allocation indices. in *Proc. of American Control Conf.*, pages 42204225, Seattle, WA, June 2008.
16. V.S.B. Konstantin E. Avrachenkov. Whittle index policy for crawling ephemeral content. in *Proc. of 2015 IEEE 54th Annual Conference on Decision and Control (CDC)*, 2015.
17. K. Liu and Q. Zhao. Indexability of restless bandit problems and optimality of whittle index for dynamic multichannel access. IEEE Trans. Inf. Theory, 56(11):5547-5567, Nov. 2010.
18. W. Ouyang, S. Murugesan, A. Eryilmaz, N.B. Shroff, Exploiting channel memory for joint estimation and scheduling in downlink networks—a Whittle's indexability analysis. IEEE Trans. Inf. Theory **4**(61), 1702–1719 (2015)
19. F. Cecchi, P. Jacko. Scheduling of users with markovian time-variyng transmission rates. in *Proc. ACM Sigmetrics*, Pittsburgh, PA, June 2013.

20. S. Aalto, P. Lassila, P. Osti. Whittle index approach to size-aware scheduling with timevarying channels. in *Proc. ACM Sigmetrics*, Portland, OR, June 2015.
21. S. Aalto, P. Lassila, P. Osti, Whittle index approach to size-aware scheduling for timevarying channels with multiple states. Queueing Syst. **83**, 195–225 (2016)
22. J. Gittins, K. Glazebrook, R. Weber, *Multi-armed Bandit Allocation Indices*, 2nd edn. (John Wiley & Sons Ltd., New York, NY, 2011)
23. D.R. Hernandez, *Indexable Restless Bandits* (VDM Verlag, 2008)
24. P. Jacko, *Dynamic Priority Allocation in Restless Bandit Models* (LAP Lambert Academic Publishing, 2010)
25. T.J.S.H. Ahmad, M. Liu, Q. Zhao, B. Krishnamachari, Optimality of myopic sensing in multi-channel opportunistic access. IEEE Trans. Inf. Theory **55**(9), 4040–4050 (2009)
26. S. Ahmad and M. Liu. Multi-channel opportunistic access: a case of restless bandits with multiple players. in *Proc. Allerton Conf. Commun. Control Comput*, Oct. 2009, pp. 1361–1368.

Chapter 4
Myopic Policy for Opportunistic Scheduling: Homogeneous Multistate Channels

4.1 Introduction

4.1.1 Background

Consider a downlink scheduling wireless communication system composed of one user and N homogeneously independent channels, each of which is modeled as an X-state Markov chain with known matrix of transition probabilities. At each time period one channel is scheduled to work and some reward depending on the state of the worked channel is obtained. The objective is to design a scheduling policy maximizing the expected accumulated discounted reward over an infinite time horizon. Mathematically, the scheduling problem can be formulated into a RMAB problem in decision theory [1]. Actually, RMAB has a wide range of applications, i.e., wired/wireless communication systems, manufacturing systems, transportation systems, economic systems, statistics, biomedical engineering, and information systems [1, 2]. However, the RMAB problem is proved to be PSPACE-Hard [3].

Specifically, the challenges of multistate RMAB are threefold: First, the probability vector is not completely ordered in probability space, making structural analysis substantially more difficult; Second, multistate RMAB tends to encounter the "curse of dimensionality," which is further complicated by the uncountably infinite probability space; Third, the imperfect observation brings out nonlinear propagation of belief information about system state. Although the first two factors are usually taken into account, there are few literature in RMAB involving nonlinear belief information propagation. Hence, numerical approach is used to evaluate policy performance generally. In fact, the numerical approach has huge computational complexity and cannot provide any meaningful insight into optimal policy.

K. Wang, L. Chen, *Restless Multi-Armed Bandit in Opportunistic Scheduling*,
https://doi.org/10.1007/978-3-030-69959-8_4

For the three reasons, in this paper, we study two instances of the generic RMAB in which the optimal policy is the so-called myopic policy or greedy policy with linear computation complexity $O(N)$. Specially, we develop two sets of sufficient conditions to guarantee the optimality of the myopic policy for the two instances. In a nutsell, the optimal policy is to schedule the best channels each time in the sense of MLR order [4].

4.1.2 Existing Work

In the generic description of RMABs, there are N arms evolving as Markov chains. A player activates multiple arms out of N arms each time, and receives a reward depending on the states of the activated arms. The objective of the play is to maximize the expected reward sum over an infinite horizon by deciding which arms should be activated each time. As a simple instance, if the unactivated arms hold their states, then the RMAB problem is degenerated into the MAB one which was solved to reveal that the optimal policy has an index structure [5, 6].

As far as RMAB problem is concerned, there are two major research lines in this field. The first research thrust is to analyze performance gap between optimal policy and approximation policy [7–9]. Specifically, a simple myopic policy was shown to achieve a factor 2 approximation of optimal policy for a subclass of *Monotonic* MAB [8, 9]. The second direction is to seek sufficient condition to guarantee the optimality of the myopic policy for some instances of RMAB, particularly in the scheduling field of opportunistic communications [10–17].

To characterize the "myopic" (or "greedy") feature, we need a certain kind of order concerning the information states of all channels, which makes the multistate channel model different from two-state one to a large extent. Specifically, the optimality of the myopic policy requires total order for two-state model while requires partial order for multistate one.

For *two-state* Markov channel model, the structure and partial optimality of the myopic policy were obtained for homogeneous Markov channels in [17]. Then the optimality of the myopic sensing policy was obtained for the positively correlated homogeneous Markov channels for accessing one channel in [18], and further was extended to access multiple homogeneous channels in [10]. From the viewpoint of exploitation dominating exploration for myopic policy, in [15], we extended homogeneous channels to heterogeneous ones, and derived a set of closed-form sufficient conditions to guarantee the optimality of myopic sensing policy. The authors [12] studied the myopic channel probing policy for the scenario with imperfect observation, but only established its optimality in the particular case of probing one channel each time. However, we established the optimality of myopic policy for the case of probing $N-1$ of N channels each time, analyzed the performance of the myopic probing policy by domination theory in [16]. Further, we studied accessing arbitrary number of channels each time and derived more generic conditions on the optimality in [14].

For *multistate* Markov channel model [19–21], the authors in [19] established sufficient conditions for optimality of myopic sensing policy (in sense of first-order stochastic dominance (FOSD)) in multistate homogeneous channels with perfect observation. For the same channel model, we study the optimality of myopic policy when transition matrix has a non-trivial eigenvalue with N -1 times [21]. Actually, due to noise, detecting methods, and etc., it is hard to implement the perfect state observation in practical applications.

4.1.3 Main Results and Contributions

Hence, in this paper, we consider the problem of imperfect (or indirect) observation of channel states, and analyze its impact on scheduling channels, which makes our scheduling problem different from [19, 21] to a large extent. Under indirect or imperfect observation, an observation matrix is introduced, from the viewpoint of mathematics, to replace the identity matrix for the direct observation case considered in [19, 21]. Hence, the FOSD order adopted in [19, 21] is not sufficient to characterize the order of information states for imperfect observation. Thus, the MLR order, a kind of stronger stochastic order than FOSD, is utilized to describe the order structure of information states. Moreover, the basic approach used in this paper is totally different from [19] in deriving the optimality of myopic policy. Specifically, our argument is more generic, and can be extended to the case of *heterogeneous* multistate model. As a result, the sufficient conditions obtained in this paper cannot degenerate into those of [19] for the perfect observation, which demonstrates the tradeoff between more generic method and stricter condition.

In particular, the contributions of this paper include the following:

- The structure of the myopic policy is shown to be a simple queue determined by the information states of channels provided that certain conditions are satisfied for the observation matrix and the probability transition matrix of homogeneous multistate channels.
- We establish two sets of conditions to ensure the optimality of myopic policy for two different scenarios—one is the case of "positive" order on the row vectors of probability transition matrix, and another is the "inverse" order on the row vectors of that matrix.
- Our derivation demonstrates the advantage of branch-and-bound and the directed comparison-based optimization approach.

Notation: Boldface lower and upper case letters represent column vectors and matrices, respectively.

Table 4.1 Main notations

Symbols	Descriptions
$\mathcal{N}$	The set of N channels, i.e., 1, 2, $\cdots$, N
$\mathcal{X}$	The set of X states, i.e., 1,2,. . .X
$\mathbf{A}$	Channel state transition matrix
$\boldsymbol{B}$	Observation probability matrix
$a_{i,\,j}$	The transition probability of from state i to j
b_{im}	The probability of being observed as state m when state i
$s_t^{(n)}$	The state of channel n in slot t
$x_0^{(n)}$	Initial probability distribution of channel n
$\mathbf{x}_t^{(n)}$	The information state at time t
$\mathbf{r}$	The X-dimensional reward column vector
u_t	The allocation policy in time t
T	The total number of time slots
t	Time slot index
$\mathbf{e}_i$	An X-dimensional column vector with 1 in the i-th element and 0 in others
$\mathbf{E}$	The $X \times X$ identity matrix
$[\mathbf{A}]_k$, $[\mathbf{A}]_{:,\,l}$	The kth row, the lth column, respectively
$[\mathbf{A}]_{k,\,l}$	The element in the kth row and the lth column of matrix A
$(\cdot)^*$	The complex conjugate
$\mathrm{tr}(\cdot)$, $(\cdot)^{\mathrm{T}}$	The trace, transpose, of a matrix, respectively
$\mathrm{diag}(a_1,\ldots,a_K)$	A diagonal and a block-diagonal matrix with $a_1 \ldots a_K$

4.2 Problem Formulation

4.2.1 System Model

Table 4.1 summaries main notations used in this chapter.

Consider a wireless downlink slotted system consisting of one user and N independent channels $n = 1, \cdots, N$. Assume each channel n has a finite number, X, of states, denoted as $\mathcal{X} = \{1, 2, \cdots, X\}$. Let $s_t^{(n)}$ denote the state of channel n at time slot $t = 1, 2, \cdots$. The state $s_t^{(n)}$ evolves according to an X-state homogeneous Markov chain with transition probability matrix $\mathbf{A} = (a_{ij})_{i,j\in\mathcal{X}}$, where,

$$a_{ij} = \mathbb{P}\left(s_{t+1}^{(n)} = j \mid s_t^{(n)} = i\right)$$

All channels are initialized with $s_0^{(n)} \sim x_0^{(n)}$, where $x_0^{(n)}$ are specified initial distributions for $n = 1, \cdots, N$.

At each time instant t, only one of these channels can be allocated to the user. If channel n is allocated at time t, an instantaneous reward $\beta^t r\left(s_t^{(n)}\right)$ is accrued, herein, $0 \leqslant \beta \leqslant 1$ denotes the discount factor.

After transmitting information over the chosen channel n, the state of channel n is indirectly observed via noisy measurement (i.e., feedback information) to be $y_{t+1}^{(n)}$ of the channel state $s_{t+1}^{(n)}$. Assume that these observations $y_{t+1}^{(n)}$ belong to a finite set $\mathcal{Y}$ indexed by $m = 1, \cdots, Y$. Let $\mathbf{B} = (b_{im})_{i \in X, m \in \mathcal{Y}}$ denote the homogeneous observation probability matrix, where each element is $b_{im} \triangleq \mathbb{P}\left(y_{t+1}^{(n)} = m \mid s_t^{(n)} = i, u_t = n\right)$. That is, $\mathbf{B}$ is homogeneous.

Let $u_t \in \{1, \cdots, N\}$ denote which channel is allocated at time t. Consequently, $s_{t+1}^{(u_t)}$ denotes the state of the allocated channel at time $t + 1$. Denote the observation history at time t as $Y_t := \left(y_1^{(u_0)}, \cdots, y_t^{(u_{t-1})}\right)$ and the decision history as $U_t := (u_0, \cdots, u_t)$. Then the channel at time $t + 1$ is chosen according to $u_{t+1} = \mu(Y_{t+1}, U_t)$, where the policy denoted as μ belongs to the class of stationary policies $\mathcal{U}$. The total expected discounted reward over an infinite time horizon is given by

$$J_\mu = \mathbb{E}\left[\sum_{t=0}^{\infty} \beta^t r\left(s_t^{(u_t)}\right)\right], \quad u_t = \mu(Y_t, U_{t-1}) \tag{4.1}$$

where $\mathbb{E}$ denotes mathematical expectation.

The aim is to determine the optimal stationary policy

$$\mu^* = \underset{\mu \in \mathcal{U}}{\operatorname{argmax}} J_\mu \tag{4.2}$$

which yields the maximum rewards $J^* = J_{\mu^*}$ in (4.1).

4.2.2 Information State

The above partially observed multiarmed bandit problem can be re-expressed as a fully observed multiarmed bandit in terms of the information state. For each channel n, denoted by $\mathbf{x}_t^{(n)}$ the information state at time t (Bayesian posterior distribution of $s_t^{(n)}$) as

$$\mathbf{x}_t^{(n)} = \left(x_t^{(n)}(1), x_t^{(n)}(2), \cdots, x_t^{(n)}(X)\right)$$

where $x_t^{(n)}(i) \triangleq \mathbb{P}\left(s_t^{(n)} = i \mid Y_t, U_{t-1}\right)$.

The hidden markovian model (HMM) multiarmed bandit problem can be viewed as the following scheduling problem: Consider N parallel HMM state estimation filters, one for each channel. The channel n is allocated, an observation $y_{t+1}^{(n)}$ is obtained and the information state $\mathbf{x}_{t+1}^{(n)}$ is computed recursively by the HMM state filter according to

$$\mathbf{x}_{t+1}^{(n)} = \Gamma\left(\mathbf{x}_t^{(n)}, y_{t+1}^{(n)}\right), \text{if channel } n \text{ is allocated at time } t \tag{4.3}$$

where

$$\Gamma\left(\mathbf{x}^{(n)}, y^{(n)}\right) \triangleq \frac{\mathbf{B}\left(y^{(n)}\right)\mathbf{A}^{\top}\mathbf{x}^{(n)}}{d(\mathbf{x}^{(n)}, y^{(n)})}, \tag{4.4}$$

$$d\left(\mathbf{x}^{(n)}, y^{(n)}\right) \triangleq 1_X^{\top}\mathbf{B}\left(y^{(n)}\right)\mathbf{A}^{\top}\mathbf{x}^{(n)}.$$

In (4.4), if $y^{(n)} = m$, then $\mathbf{B}(m) = \ diag\ [b_{1m}, \cdots, b_{Xm}]$ is the diagonal matrix formed by the mth column of the observation matrix $\mathbf{B}$, $\mathbf{A}_i$ is the ith row of the matrix $\mathbf{A}$, $\mathbf{1}_X$ is an X-dimensional column vector of ones, and $\mathbf{E}$ is an identity matrix.

The state estimation of the other $N-1$ channels is according to

$$\mathbf{x}_{t+1}^{(l)} = \mathbf{A}^{\top}\mathbf{x}_t^{(l)} \tag{4.5}$$

if channel l is not allocated at time t, $l \in \{1, \cdots, N\}, \quad l \neq n$.

Let $\Pi(X)$ denote the state space of information states $\mathbf{x}^{(n)}$, $n \in \{1, 2, \cdots, N\}$, which is a $X-1$-dimensional simplex:

$$\Pi(X) = \{\mathbf{x} \in \mathbb{R}^X : \mathbf{1}_X^{\top}\mathbf{x} = 1, 0 \leqslant x(i) \leqslant 1 \text{ for all } i \in \mathcal{X}\} \tag{4.6}$$

The process $\mathbf{x}_t^{(n)}, n = 1, \cdots, N$ qualifies as an information state since choosing $u_{t+1} = \mu(Y_{t+1}, U_t)$ is equivalent to choosing $u_{t+1} = \mu\left(\mathbf{x}_{t+1}^{(1)}, \cdots, \mathbf{x}_{t+1}^{(N)}\right)$. Using the smoothing property of conditional expectations, the reward function (4.1) can be rewritten in terms of the information state as

$$J_{\mu} = \mathbb{E}\left[\sum_{t=0}^{\infty} \beta^t \mathbf{r}^{\top}\mathbf{x}_t^{(u_t)}\right], \quad u_t = \mu\left(\mathbf{x}_t^{(1)}, \cdots, \mathbf{x}_t^{(N)}\right) \tag{4.7}$$

where $\mathbf{r}$ denotes the X-dimensional reward column vector with $r(1) \leqslant r(2) \leqslant \cdots \leqslant r(X)$. The aim is to compute the optimal policy $\underset{\mu \in \mathcal{U}}{\operatorname{argmax}} J_{\mu}$.

For convenient analysis of the optimization problem in (4.8), we derive its dynamic programming formulation for a finite time T as follows:

$$\begin{cases} V_T\left(\mathbf{x}_T^{(1:N)}\right) &= \max_{u_T} \mathbb{E}\left[\mathbf{r}^\top \mathbf{x}_T^{(u_T)}\right] \\ V_t\left(\mathbf{x}_t^{(1:N)}\right) &= \max_{u_t} \mathbb{E}\left[\mathbf{r}^\top \mathbf{x}_t^{(u_t)} + \beta \sum_{m\in\mathcal{Y}} d\left(\mathbf{x}_t^{(u_t)}, m\right) V_{t+1}\left(\mathbf{x}_{t+1}^{(1:u_t-1)}, \mathbf{x}_{t+1,m}^{(u_t)}, \mathbf{x}_{t+1}^{(u_t+1:N)}\right)\right] \end{cases} \tag{4.8}$$

where, $\mathbf{x}_t^{(i:j)} \triangleq \left(\mathbf{x}_t^{(i)}, \mathbf{x}_t^{(i+1)}, \cdots, \mathbf{x}_t^{(j)}\right)$, and

$$\begin{cases} \mathbf{x}_{t+1,m}^{(u_t)} = \Gamma\left(\mathbf{x}_t^{(u_t)}, m\right) \\ \mathbf{x}_{t+1}^{(n)} = \mathbf{A}^\top \mathbf{x}_t^{(n)}, n \neq u_t \end{cases} \tag{4.9}$$

When $T \to \infty$, $J^* = V_0(\mathbf{X}^{(1\,:\,N)})$.

4.2.3 Myopic Policy

Theoretically, we can solve the above dynamic programming by backward deduction to obtain optimal policy. However, obtaining the optimal solution directly from the above recursive equations is computationally prohibitive. Hence, we turn to seek a simple myopic policy only maximizing the immediate reward, which is easy to compute and implement, formally defined as follows:

$$\widehat{u}_t = \underset{u_t}{\operatorname{argmax}}\, \mathbf{r}^\top \mathbf{x}_t^{(u_t)} \tag{4.10}$$

As we know, the myopic policy or greedy policy is not optimal generally, but its computation complexity is very low in $O(N)$. Thus, our aim in the following parts is to seek sufficient conditions to guarantee that the myopic policy is optimal.

Considering the multivariate information state, we introduce some partial orders to characterize the "order" of information states in the following sections.

Definition 4.1 (MLR ordering [4]) Let $\mathbf{x}_1$, $\mathbf{x}_2 \in \Pi(X)$ be any two belief vectors. Then $\mathbf{X}_1$ is greater than $\mathbf{X}_2$ with respect to the MLR ordering—denoted as $\mathbf{x}_1 \geqslant_r \mathbf{x}_2$, if

$$x_1(i)x_2(j) \leqslant x_2(i)x_1(j), i < j, i, j \in \{1, 2, \ \ldots, X\}.$$

Definition 4.2 (FOSD ordering [4]) Let $\mathbf{x}_1$, $\mathbf{x}_2 \in \Pi(X)$ then $\mathbf{x}_1$ first order stochastically dominates $\mathbf{x}_2$—denoted as $x_1 \geqslant_s x_2$, if the following exists for $j = 1, 2, \cdots, X$,

$$\sum_{i=j}^{X} x_1(i) \geqslant \sum_{i=j}^{X} x_2(i)$$

Some useful results for MLR and FOSD order [4] are stated here:

Proposition 4.1 ([4]) *Let* $\mathbf{x}_1$, $\mathbf{x}_2 \in \Pi(X)$, *the following holds*

(i) $\mathbf{x}_1 \geqslant_r \mathbf{x}_2$ *implies* $\mathbf{x}_1 \geqslant_s \mathbf{x}_2$.
(ii) *Let* $\mathcal{V}$ *denote the set of all X-dimensional vectors* $\mathbf{v}$ *with nondecreasing components, i.e.,* $v_1 \leqslant v_2 \leqslant \cdots \leqslant v_X$ *Then* $\mathbf{X}_1 \geqslant_s \mathbf{X}_2$ *i.f.f. for all* $\mathbf{v} \in \mathcal{V}, \mathbf{v}^\top \mathbf{x}_1 \geqslant \mathbf{v}^\top \mathbf{x}_2$.

Based on partial order structure, we now describe the structure of the myopic policy.

Proposition 4.2 (Structure of Myopic Policy) *If,* $\mathbf{x}_t^{(\sigma_1)} \geqslant_r \mathbf{x}_t^{(\sigma_2)} \geqslant_r \cdots \geqslant_r \mathbf{x}_t^{(\sigma_N)}$, *herein,* $\{\sigma_1, \sigma_2, \cdots, \sigma_N\}$ *is a permutation of* $\{1, \cdots, N\}$, *then the myopic policy at t is*

$$\widehat{u}_t = \mu_t\left(\mathbf{x}_t^{(1)}, \cdots, \mathbf{x}_t^{(N)}\right) = \sigma_1 \tag{4.11}$$

4.3 Optimality Analysis of Myopic Policy

To analyze the performance of the myopic policy, we first introduce an auxiliary value function and then prove a critical feature of the auxiliary value function. Next, we give some assumptions about transition matrix, and show its special stochastic order. Finally, by deriving the bounds of different policies, we get some important bounds, which serve as the basis to prove the optimality of the myopic policy.

4.3.1 Value Function and Its Properties

First, we define the auxiliary value function (AVF) as follows:

$$\begin{cases} W_T^{(\widehat{u_T})}\left(\mathbf{x}_T^{(1:N)}\right) = \mathbf{r}^\top \mathbf{x}_T^{(\widehat{u_T})}, \\ W_\tau^{(\widehat{u_\tau})}\left(\mathbf{x}_\tau^{(1:N)}\right) = \mathbf{r}^\top \mathbf{x}_\tau^{(\widehat{u_\tau})} + \underbrace{\beta \sum_{m\in\mathcal{Y}} d\left(\mathbf{x}_\tau^{(\widehat{u_\tau})}, m\right) W_{\tau+1}^{(\widehat{u_{\tau+1}})}\left(\mathbf{x}_{\tau+1}^{(1:\widehat{u_\tau}-1)}, \mathbf{x}_{\tau+1,m}^{(\widehat{u_\tau})}, \mathbf{X}_{\tau+1}^{(\widehat{u_\tau}+1:N)}\right)}_{F(\mathbf{x}_T^{(1:N)}, \widehat{u_\tau})}, \\ W_t^{(u_t)}\left(\mathbf{x}_t^{(1:N)}\right) = \mathbf{r}^\top \mathbf{x}_t^{(u_t)} + \underbrace{\beta \sum_{m\in\mathcal{Y}} d\left(\mathbf{x}_t^{(u_t)}, m\right) W_{t+1}^{(\widehat{u_{t+1}})}\left(\mathbf{x}_{t+1}^{(1:u_t-1)}, \mathbf{x}_{t+1,m}^{(u_t)}, \mathbf{x}_{t+1}^{(u_t+1:N)}\right)}_{F(\mathbf{x}_t^{(1:N)}, u_t)} \end{cases} \tag{4.12}$$

where, $t + 1 \leqslant \tau \leqslant T$.

Remark 4.1 AVF is the reward under the policy: at slot t, ut is adopted, while after t, myopic policy $\widehat{u}_\tau (t + 1 \leqslant \tau \leqslant T)$ is adopted.

Lemma 4.1 $W_t^{(u_t)}\left(\mathbf{x}_t^{(1:N)}\right)$ *is decomposable for all* $t = 0, 1, \cdots, T$, i.e.,

$$\begin{aligned} W_t^{(u_t)}\left(\mathbf{x}_t^{(1:n-1)}, \mathbf{x}_t^{(n)}, \mathbf{x}_t^{(n+1:N)}\right) &= \sum_{i=1}^{X} x_t^{(n)}(i) W_t^u\left(\mathbf{x}_t^{(1:n-1)}, \mathbf{e}_i, \mathbf{x}_t^{(n+1:N)}\right) \\ &= \sum_{i=1}^{X} \mathbf{e}_i^\top \mathbf{x}_t^{(n)} W_t^{(u_t)}\left(\mathbf{x}_t^{(1:n-1)}, \mathbf{e}_i, \mathbf{x}_t^{(n+1:N)}\right) \end{aligned}$$

Proof Please refer to Appendix 4.6.1. ■

4.3.2 Assumptions

We make the following assumptions/conditions.

Assumption 1 *Assume that*

(i) $\mathbf{A}_1 \leq_r \mathbf{A}_2 \leq_r \ldots \leq_r A_X$.
(ii) $\mathbf{B}_{:,1} \leq_r \mathbf{B}_{:,2} \leq_r \ldots \leq_r \mathbf{B}_{:,Y}$.
(iii) *There exists some* $K (2 \leqslant K \leqslant Y)$ *such that*

$$\Gamma(\mathbf{A}^\top \mathbf{e}_1, K) \geqslant_r (\mathbf{A}^\top)^2 \mathbf{e}_X$$

$$\Gamma(\mathbf{A}^\top \mathbf{e}_X, K-1) \leqslant_r (\mathbf{A}^\top)^2 \mathbf{e}_1$$

(iv) $\mathrm{A}_1 \leqslant_r \mathbf{x}_0^{(1)} \leqslant_r \mathbf{x}_0^{(2)} \leqslant_r \cdots \leqslant_r \mathbf{x}_0^{(N)} \leqslant_r \mathbf{A}_X$

(v) $\mathbf{r}^{\top}(\mathbf{e}_{i+1} - \mathbf{e}_i) \geqslant \mathbf{r}^{\top}\mathbf{Q}^{\top}(\mathbf{e}_{i+1} - \mathbf{e}_i)(1 \leqslant i \leqslant X - 1)$, where $\mathbf{A} = \mathbf{V}\Lambda\mathbf{V}^{-1}$, $\mathbf{Q} = \mathbf{V}\Upsilon\mathbf{V}^{-1}$

$$\Lambda = \begin{pmatrix} 1 & 0 & \dots & 0 \\ 0 & \lambda_2 & \dots & 0 \\ \vdots & \vdots & \ddots & \vdots \\ 0 & 0 & \dots & \lambda_X \end{pmatrix}, \Upsilon = \begin{pmatrix} 1 & 0 & \dots & 0 \\ 0 & \frac{\beta\lambda_2}{1-\beta\lambda_2} & \dots & 0 \\ \vdots & \vdots & \ddots & \vdots \\ 0 & 0 & \dots & \frac{\beta\lambda_X}{1-\beta\lambda_X} \end{pmatrix}$$

Remark 4.2 Assumption 1(i) ensures that the higher the quality of the channel's current state, the higher is the likelihood that the next channel state will be of high quality. Assumption 1(ii) ensures that the higher the quality of the channel's current state, the higher is the observation likelihood that the next channel state will be of high quality. Assumption 1(iii) along with 1.1–1.2 ensure that the information states of all channels can be ordered at all times in the sense of MLR order (see the proof of Proposition 4.6). Assumption 1(iv) states that initially the channels can be ordered in terms of their quality. Basically, Assumption 1(i)–(iv) ensure the order of information states, while Assumption 1(v) is required for reward comparison. In particular, Assumption 1(v) states that the instantaneous reward obtained at different states is sufficiently separated.

Example 4.1 When $X = 2$, $Y = 2$, we have $\lambda_2 = a_{22} - a_{12}$. Assumption 1 degenerates into the following [22]:

(i) $a_{22} \geqslant a_{12}$, that is, $\lambda_2 \geqslant 0$

(ii–iii) $b_{22} \geqslant b_{12}$,

(iv) $a_{12} \leqslant x_0^{(1)}(2) \leqslant \mathbf{x}_0^{(2)}(2) \leqslant \cdots \leqslant x_0^{(N)}(2) \leqslant a_{22}$,

(v) $r(2) - r(1) \geqslant \frac{\beta\lambda}{1-\beta\lambda}(r(2) - r(1))$, that is, $\beta\lambda_2 \leqslant \frac{1}{2}$

4.3.3 Properties

Under Assumption 1(i)–(v), we have some important propositions concerning the structure of information state in the following, which are proved in Appendix A.

Proposition 4.3 *Let* $\mathbf{x}_1, \mathbf{x}_2 \in \Pi(X)$ *and* $\mathbf{x}_1 \leqslant \mathbf{x}_2$, *then*

$$(\mathbf{A}_1)^{\mathrm{T}} \leq_r \mathbf{A}^{\mathrm{T}}\mathbf{x}_1 \leq_r \mathbf{A}^{\mathrm{T}}\mathbf{x}_2 \leq_r (\mathbf{A}_X)^{\mathrm{T}}.$$

Proof Suppose $i > j$, we have

$$(\mathbf{e}_i^\top \mathbf{A}^\top \mathbf{x}_2) \cdot (\mathbf{e}_j^\top \mathbf{A}^\top \mathbf{x}_1) - (\mathbf{e}_j^\top \mathbf{A}^\top \mathbf{x}_2) \cdot (\mathbf{e}_i^\top \mathbf{A}^\top \mathbf{x}_1)$$

$$= \sum_{k=1}^{X} \mathbf{A}_{k,i} x_2(k) \sum_{l=1}^{X} \mathbf{A}_{l,j} x_1(l) - \sum_{k=1}^{X} \mathbf{A}_{k,j} x_2(k) \sum_{l=1}^{X} \mathbf{A}_{l,i} x_1(l)$$

$$= \sum_{l=1}^{X} \sum_{k=l}^{X} \left(\mathbf{A}_{k,i}\mathbf{A}_{l,j} - \mathbf{A}_{l,i}\mathbf{A}_{k,j}\right)\left(x_2(k)x_1(l) - x_2(l)x_1(k)\right) \geqslant 0$$

where the last inequality is due to $\mathbf{A}_k \geqslant_r \mathbf{A}_l (k \geqslant l)$ and $\mathbf{x}_2 \geqslant_r \mathbf{x}_1$. ∎

Then, considering $\mathrm{e}_1 \leqslant_r \mathbf{x}_1 \leqslant_r \mathbf{x}_2 \leqslant_r e_X$, we have

$$(\mathbf{A}_1)^\top = \mathbf{A}^\top \mathbf{e}_1 \leqslant_r \mathbf{A}^\top \mathbf{x}_1 \leqslant_r \mathbf{A}^\top \mathbf{x}_2 \leqslant_r \mathbf{A}^\top \mathbf{e}_X = (\mathbf{A}_X)^\top$$

Proposition 4.3 states that if at any time t the information states of two channels are stochastically ordered and none of these channels is chosen at t, then the same stochastic order between the information states at time $t + 1$ is maintained.

Proposition 4.4 *Let* $\mathbf{x}_1, \mathbf{x}_2 \in \Pi(X)$ *and* $(\mathbf{A}_1)^\top \leqslant_r \mathbf{X}_1 \leqslant_r \mathbf{X}_2 \leqslant_r (\mathbf{A}_X)^\top$, *then for* $1 \leqslant k \leqslant Y$

$$\Gamma(\mathbf{x}_1, k) \leqslant_r \Gamma(\mathbf{x}_2, k)$$

Proof According to Proposition 4.3, we have $\mathbf{z}_1 = \mathbf{A}^\top \mathbf{x}_1 \leqslant_r \mathbf{A}^\top \mathbf{x}_2 = \mathbf{z}_2$. Suppose $i > j$, we have

$$(\Gamma(\mathbf{x}_2, k))_i \cdot (\Gamma(\mathbf{x}_1, k))_j - (\Gamma(\mathbf{x}_2, k))_j \cdot (\Gamma(\mathbf{x}_1, k))_i$$

$$= \frac{b_{ik} z_2(i)}{\sum_{x=1}^{X} b_{xk} z_2(x)} \cdot \frac{b_{jk} z_1(j)}{\sum_{x=1}^{X} b_{xk} z_1(x)} - \frac{b_{jk} z_2(j)}{\sum_{x=1}^{X} b_{xk} z_2(x)} \cdot \frac{b_{ik} z_1(i)}{\sum_{x=1}^{X} b_{xk} z_1(x)}$$

$$= \frac{b_{ik} b_{jk} (z_2(i) z_1(j) - z_2(j) z_1(i))}{\sum_{x=1}^{X} b_{xk} z_2(x) \sum_{x=1}^{X} b_{xk} z_1(x)} \geqslant 0$$

where, $z_2(i)z_1(j) - z_2(j)z_1(i) \geqslant 0$ is from $z_1 \leqslant_r \mathbf{z}_2$. ∎

Proposition 4.4 states the increasing monotonicity of updating rule with information state for scheduled channel.

Proposition 4.5 *Let* $\mathbf{x} \in \Pi(X)$ *and* $(\mathbf{A}_1)^\top \leqslant_r \mathbf{X} \leqslant_r (\mathbf{A}_X)^\top$, *then* $\Gamma(\mathbf{x}, k) \leqslant_r \Gamma(\mathbf{x}, m)$ *for any* $1 \leqslant k \leqslant m \leqslant Y$.

Proof Let $\mathbf{z} = \mathbf{A}^\top x$. Suppose $i > j$, we have

$$
\begin{aligned}
&(\Gamma(\mathbf{x},m))_i \cdot (\Gamma(\mathbf{x},k))_j - (\Gamma(\mathbf{x},m))_j \cdot (\Gamma(\mathbf{x},k))_i \\
&= \frac{b_{im}z(i)}{\sum_{l=1}^{X} b_{lm}z(l)} \cdot \frac{b_{jk}z(j)}{\sum_{l=1}^{X} b_{lk}z(l)} - \frac{b_{jm}z(j)}{\sum_{l=1}^{X} b_{lm}z(l)} \cdot \frac{b_{ik}z(i)}{\sum_{l=1}^{X} b_{lk}z(l)} \\
&= \frac{\left(b_{im}b_{jk} - b_{jm}b_{ik}\right)z(i)z(j)}{\sum_{l=1}^{X} b_{lm}z(l) \sum_{l=1}^{X} b_{lk}z(l)} \geqslant 0
\end{aligned}
$$

where, $b_{im}b_{jk} - b_{jm}b_{ik} \geq 0$ is from $\mathbf{B}(m) \geq_r \mathbf{B}(k)$. ■

Proposition 4.5 states the increasing monotonicity of updating rule with the increasing number of observation state for scheduled channel.

Proposition 4.6 *Under Assumption* 1, *we have either* $\mathbf{x}_t^{(l)} \leqslant_r \mathbf{x}_t^{(n)}$ or $\mathbf{x}_t^{(n)} \leqslant_r \mathbf{x}_t^{(l)}$ *for all* $l, n \in \{1, 2, \cdots, N\}$ *for all t.*

Proof Based on Proposition 4.3, $\mathrm{A}^\top \mathrm{x}$ is monotonically increasing in $\mathbf{x}$ $((\mathbf{A}_1)^\top \leqslant_r \mathbf{x} \leqslant_r (\mathbf{A}_X)^\top)$. As a result, we obtain

$$
(\mathbf{A}^\top)^2 \mathbf{e}_1 = \mathbf{A}^\top (\mathbf{A}_1)^\top \leqslant \mathbf{A}^\top \mathbf{x} \leqslant \mathbf{A}^\top (\mathbf{A}_X)^\top = (\mathbf{A}^\top)^2 \mathbf{e}_X \tag{4.13}
$$

Based on Propositions 4.4 and 4.5, $\Gamma(\mathbf{x}, k)$ is monotonically increasing in both $\mathbf{x}$ $((\mathbf{A}_1)^\top \leqslant_r \mathbf{X} \leqslant_r (\mathbf{A}_X)^\top)$ and $k(1 \leqslant k \leqslant Y)$. Consequently, we have

$$
\Gamma\left((\mathbf{A}_1)^\top, 1\right) \leqslant \Gamma(\mathbf{x}, k) \leqslant \Gamma\left((\mathbf{A}_X)^\top, K-1\right), \text{for } (\mathbf{A}_1)^\top \leqslant_r \mathbf{x} \leqslant_r (\mathbf{A}_X)^\top, 1 \leqslant k < K \tag{4.14}
$$

$$
\Gamma\left((\mathbf{A}_1)^\top, K\right) \leqslant \Gamma(\mathbf{x}, k) \leqslant \Gamma\left((\mathbf{A}_X)^\top, Y\right), \quad \text{for } (\mathbf{A}_1)^\top \leqslant_r \mathbf{x} \leqslant_r (\mathbf{A}_X)^\top, K \leqslant k \leqslant Y. \tag{4.15}
$$

Combining (4.13), (4.14), (4.15), and Assumption 1(iii), we can obtain that the evolution function (4.9) of information state is monotonically increasing in x $((A_1)^\top \leqslant \mathbf{x} \leqslant (\mathbf{A}_X)^\top)$. Combining Assumption 1(iv), we conclude the proposition. ■

Proposition 4.6 states that under Assumption 1, the information states of all channels can be ordered stochastically or be comparable at all times.

Now we give an important structural property on transition matrix in the following proposition.

Proposition 4.7 *Suppose that transition matrix* $\mathbf{A}$ *has X eigenvalues* $\lambda_1 \geqslant \lambda_2 \cdots \geqslant \lambda_X$ *and the corresponding eigenvectors are* $\mathbf{v}_1, \mathbf{v}_2, \cdots, \mathbf{v}_X$. *If* $\mathbf{x}_1, \mathbf{x}_2 \in \Pi(X)$, *then we have*

(i) $\lambda_1 = 1$ and $\mathbf{v}_1 = \frac{1}{\sqrt{X}} \mathbf{1}_X$

(ii) *for any* λ,

$$\Lambda_1 \mathbf{V}^\top (\mathbf{x}_1 - \mathbf{x}_2) = \Lambda_2 \mathbf{V}^\top (\mathbf{x}_1 - \mathbf{x}_2) \tag{4.16}$$

where

$$\Lambda_1 = \begin{pmatrix} \lambda_1 & 0 & \dots & 0 \\ 0 & \lambda_2 & \dots & 0 \\ \vdots & \vdots & \ddots & \vdots \\ 0 & 0 & \dots & \lambda_X \end{pmatrix}, \Lambda_2 = \begin{pmatrix} \lambda & 0 & \dots & 0 \\ 0 & \lambda_2 & \dots & 0 \\ \vdots & \vdots & \ddots & \vdots \\ 0 & 0 & \dots & \lambda_X \end{pmatrix}$$

Proof (1) For the property of $\lambda_1 = 1$ and $\mathrm{v}_1 = \frac{1}{\sqrt{X}} 1_X$, it is easily verified, i.e.,

$$\frac{1}{\sqrt{X}} \mathbf{A} \cdot \mathbf{1}_X = \frac{1}{\sqrt{X}} \begin{pmatrix} \mathbf{A}_1 \cdot \mathbf{1}_X \\ \mathbf{A}_2 \cdot \mathbf{1}_X \\ \vdots \\ \mathbf{A}_X \cdot \mathbf{1}_X \end{pmatrix} = \frac{1}{\sqrt{X}} \begin{pmatrix} 1 \\ 1 \\ \vdots \\ 1 \end{pmatrix} = \frac{1}{\sqrt{X}} 1_X$$

(2) For the property of replacing λ_1 with any value λ, we have the LHS of (4.16):

$$\begin{aligned} & \Lambda_1 \mathbf{V}^\top (\mathbf{x}_1 - \mathbf{x}_2) \\ &= \left[\lambda_1 (\mathbf{x}_1 - \mathbf{x}_2)^\top \mathbf{v}_1, \lambda_2 (\mathbf{x}_1 - \mathbf{x}_2)^\top \mathbf{v}_2, \cdots, \lambda_X (\mathbf{x}_1 - \mathbf{x}_2)^\top \mathbf{v}_X\right]^\top \\ &= \left[\lambda_1 (\mathbf{x}_1 - \mathbf{x}_2)^\top \frac{1}{\sqrt{X}} \mathbf{1}_X, \lambda_2 (\mathbf{x}_1 - \mathbf{x}_2)^\top \mathbf{v}_2, \cdots, \lambda_X (\mathbf{x}_1 - \mathbf{x}_2)^\top \mathbf{v}_X\right]^\top \\ &= \left[0, \lambda_2 (\mathbf{x}_1 - \mathbf{x}_2)^\top \mathbf{v}_2, \cdots, \lambda_X (\mathbf{x}_1 - \mathbf{x}_2)^\top \mathbf{v}_X\right]^\top \end{aligned} \tag{4.17}$$

For the RHS of (4.16), we have

$$\begin{aligned} & \Lambda_2 \mathbf{V}^\top (\mathbf{x}_1 - \mathbf{x}_2) \\ &= \left[\lambda_1 (\mathbf{x}_1 - \mathbf{x}_2)^\top \mathbf{v}_1, \lambda_2 (\mathbf{x}_1 - \mathbf{x}_2)^\top \mathbf{v}_2, \cdots, \lambda_X (\mathbf{x}_1 - \mathbf{x}_2)^\top \mathbf{v}_X\right]^\top \\ &= \left[\lambda_1 (\mathbf{x}_1 - \mathbf{x}_2)^\top \frac{1}{\sqrt{X}} \mathbf{1}_X, \lambda_2 (\mathbf{x}_1 - \mathbf{x}_2)^\top \mathbf{v}_2, \cdots, \lambda_X (\mathbf{x}_1 - \mathbf{x}_2)^\top \mathbf{v}_X\right]^\top \\ &= \left[0, \lambda_2 (\mathbf{x}_1 - \mathbf{x}_2)^\top \mathbf{v}_2, \cdots, \lambda_X (\mathbf{x}_1 - \mathbf{x}_2)^\top \mathbf{v}_X\right]^\top \end{aligned} \tag{4.18}$$

By (4.17) and (4.18), we prove the equation (4.16). ■

Proposition 4.7 states that (1) for any transition matrix, the largest eigenvalue is 1, named as *trivial eigenvalue*, and its corresponding eigenvector is $\frac{1}{\sqrt{X}} 1_X$, named as *trivial eigenvector*; (2) for any two information states, $\mathbf{x}_1$, $\mathbf{x}_2 \in \Pi(X)$, one special equation holds where the largest eigenvalue 1 can be replaced by any value.

Proposition 4.8 *Given* $\mathbf{x}_1$, $\mathbf{x}_2 \in \Pi(X)$, *we have*

$$\mathbf{r}^\top \sum_{i=1}^{\infty} (\beta \mathbf{A}^\top)^i (\mathbf{x}_1 - \mathbf{x}_2) = \mathbf{r}^\top \mathbf{Q}^\top (\mathbf{x}_1 - \mathbf{x}_2)$$

Proof

$$\begin{aligned}
\mathbf{r}^\top \sum_{i=1}^{\infty} (\beta \mathbf{A}^\top)^i (\mathbf{x}_1 - \mathbf{x}_2) &= \mathbf{r}^\top \sum_{i=1}^{\infty} \left(\beta (\mathbf{V}^{-1})^\top \Lambda \mathbf{V}^\top\right)^i (\mathbf{x}_1 - \mathbf{x}_2) \\
&\overset{(a)}{=} \mathbf{r}^\top \sum_{i=1}^{\infty} \left(\beta (\mathbf{V}^{-1})^\top \Lambda_2 \mathbf{V}^\top\right)^i (\mathbf{x}_1 - \mathbf{x}_2) \\
&= \mathbf{r}^\top \left(\mathbf{V}^{-1}\right)^\top \sum_{j=1}^{\infty} (\beta \Lambda_2)^i \mathbf{V}^\top (\mathbf{x}_1 - \mathbf{x}_2) \\
&= \mathbf{r}^\top (\mathbf{V}^{-1})^\top \Upsilon \mathbf{V}^\top (\mathbf{x}_1 - \mathbf{x}_2) \\
&= \mathbf{r}^\top (\mathbf{V} \Upsilon \mathbf{V}^{-1})^\top (\mathbf{x}_1 - \mathbf{x}_2) \\
&= \mathbf{r}^\top \mathbf{Q}^\top (\mathbf{x}_1 - \mathbf{x}_2)
\end{aligned}$$

where, the equality (a) is due to Proposition 4.7. ■

Proposition 4.8 states that the accumulated reward difference between two different state information vectors can be simply written as a matrix form.

Proposition 4.9

$$\mathbf{r}^\top \left(\mathbf{e}_i - \mathbf{e}_j\right) \geqslant \mathbf{r}^\top \mathbf{Q}^\top \left(\mathbf{e}_i - \mathbf{e}_j\right) (1 \leqslant j < i \leqslant X)$$

Proof According to Assumption 1(v), we have $\mathbf{r}^\top(\mathbf{e}_{j+1} - \mathbf{e}_j) \geqslant \mathbf{r}^\top \mathbf{Q}^\top (\mathbf{e}_{j+1} - \mathbf{e}_j)$ $(1 \leqslant j \leqslant X - 1)$. Thus, we only need to prove $\mathbf{r}^\top(\mathbf{e}_i - \mathbf{e}_j) \geqslant \mathbf{r}^\top \mathbf{Q}^\top (\mathbf{e}_i - \mathbf{e}_j)$ for any $i > j + 1$

$$\begin{aligned}
\mathbf{r}^\top \left(\mathbf{e}_i - \mathbf{e}_j\right) - \mathbf{r}^\top \mathbf{Q}^\top \left(\mathbf{e}_i - \mathbf{e}_j\right) &= \mathbf{r}^\top \sum_{k=j}^{i-1} (\mathbf{e}_{k+1} - \mathbf{e}_k) - \mathbf{r}^\top \mathbf{Q}^\top \sum_{k=j}^{i-1} (\mathbf{e}_{k+1} - \mathbf{e}_k) \\
&= \sum_{k=j}^{i-1} \left[\mathbf{r}^\top (\mathbf{e}_{k+1} - \mathbf{e}_k) - \mathbf{r}^\top \mathbf{Q}^\top (\mathbf{e}_{k+1} - \mathbf{e}_k)\right] \geqslant 0.
\end{aligned}$$

where, the last inequality is from Assumption 1(v). ■

4.3.4 *Analysis of Optimality*

We first give some bounds of performance difference on reward pairs of policies, and then derive the main theorem on the optimality of myopic policy.

Lemma 4.2 *Under Assumption* 1, $\mathbf{x}_t^l = \left(\mathbf{x}_t^{(-l)}, \mathbf{x}_t^{(l)}\right), \check{\mathbf{x}}_t^l = \left(\mathbf{x}_t^{(-l)}, \mathbf{x}_t^{(l)}\right), \mathbf{x}_t^{(l)} \leqslant_r \check{\mathbf{x}}_t^{(l)}$ (l) *we have for* $1 \leq t \leq T$

$(C1)$ if $u_t' = u_t = l$

$$\mathbf{r}^\top(\check{\mathbf{x}}_t^{(l)} - \mathbf{x}_t^{(l)}) \leq W_t^{(u_t')}(\check{\mathbf{x}}_t^l) - W_t^{(u_t)}(\mathbf{x}_t^l) \leq \sum_{i=0}^{T-t} \beta^i \mathbf{r}^\top (\mathbf{A}^\top)^i (\check{\mathbf{x}}_t^{(l)} - \mathbf{x}_t^{(l)})$$

(C2) if $u_t' \neq l, u_t \neq l$, and $u_t' = u_t$

$$0 \leq W_t^{(u_t')}(\check{\mathbf{x}}_t^l) - W_t^{(u_t)}(\mathbf{x}_t^l) \leq \sum_{i=1}^{T-t} \beta^i \mathbf{r}^\top (\mathbf{A}^\top)^i (\check{\mathbf{x}}_t^{(l)} - \mathbf{x}_t^{(l)})$$

(C3) if $u_t' = l$ and $u_t \neq l$,

$$0 \leq W_t^{(u_t')}(\check{x}_t^l) - W_t^{(u_t)}(\mathbf{x}_t^l) \leq \sum_{i=0}^{T-t} \beta^i \mathbf{r}^\top (\mathbf{A}^\top)^i (\check{\mathbf{x}}_t^{(l)} - \mathbf{x}_t^{(l)}).$$

Proof Please refer to Appendix A. ■

Remark 4.3 We would like to emphasize on what conditions the bounds of Lemma 4.2 are achieved. For (C1), the lower bound is achieved when channel l is scheduled at slot t but never scheduled after t; the upper bound is achieved when l is scheduled from t to T. For (C2), the lower bound is achieved when channel l is never scheduled from t; the upper bound is achieved when l is scheduled from $t + 1$ to T. For (C3), the lower bound is achieved when channel l is never scheduled from t; the upper bound is achieved when l is scheduled from t to T.

Lemma 4.3 *Under Assumption* 1, if $\mathbf{x}_t^{(l)} >_r \mathbf{x}_t^{(n)}$, we have $W_t^{(l)}\left(\mathbf{x}_t^{(1:N)}\right) > W_t^{(n)}\left(\mathbf{x}_t^{(1:N)}\right)$.

Proof By Lemma 4.2, we have

$$W_t^{(l)}\left(\mathbf{x}_t^{(1:N)}\right) - W_t^{(n)}\left(\mathbf{x}_t^{(1:N)}\right)$$

$$= [W_t^{(l)}(\mathbf{x}_t^{(-l)}, \mathbf{x}_t^{(l)}) - W_t^{(l)}(\mathbf{x}_t^{(-l)}, \mathbf{x}_t^{(n)})] - [W_t^{(l)}(\mathbf{x}_t^{(-l)}, \mathbf{x}_t^{(n)}) - W_t^{(n)}(\mathbf{x}_t^{(-n)}, \mathbf{x}_t^{(n)})]$$

$$\overset{(a)}{=} \left[W_t^{(l)}\left(\mathbf{x}_t^{(-l)}, \mathbf{x}_t^{(l)}\right) - W_t^{(l)}\left(\mathbf{x}_t^{(-l)}, \mathbf{x}_t^{(n)}\right)\right] - \left[W_t^{(n)}\left(\mathbf{x}_t^{(-l)}, \mathbf{x}_t^{(n)}\right) - W_t^{(n)}\left(\mathbf{x}_t^{(-n)}, \mathbf{x}_t^{(n)}\right)\right]$$

$$\geqslant \mathbf{r}^\top\left(x_t^{(l)} - \mathbf{x}_t^{(n)}\right) - \sum_{i=1}^{T-t} \beta^i \mathbf{r}^\top (\mathbf{A}^\top)^i \left(\mathbf{x}_t^{(l)} - \mathbf{x}_t^{(n)}\right)$$

$$= \mathbf{r}^\top \left(\mathbf{E} - \sum_{i=1}^{T-t} (\beta \mathbf{A}^\top)^i\right)\left(\mathbf{x}_t^{(l)} - \mathbf{x}_t^{(n)}\right)$$

$$\geqslant \mathbf{r}^\top \left(\mathbf{E} - \sum_{i=1}^{\infty} (\beta \mathbf{A}^\top)^i\right)\left(\mathbf{x}_t^{(l)} - \mathbf{x}_t^{(n)}\right)$$

$$\overset{(b)}{=} \mathbf{r}^\top (\mathbf{E} - \mathbf{V}\Upsilon\mathbf{V}^{-1})(\mathbf{x}_t^{(l)} - \mathbf{x}_t^{(n)})$$

$$= \sum_{j=2}^{X}\left[\sum_{i=j}^{X}\left(x_t^{(l)}(i) - x_t^{(n)}(i)\right)\mathbf{r}^\top(\mathbf{E} - \mathbf{Q}^\top)(\mathbf{e}_j - \mathbf{e}_{j-1})\right]$$

$$= \sum_{j=2}^{X}\left[\sum_{i=j}^{X}\left(x_t^{(l)}(i) - x_t^{(n)}(i)\right)\left[\mathbf{r}^\top(\mathbf{e}_j - \mathbf{e}_{j-1}) - \mathbf{r}^\top\mathbf{Q}^\top(\mathbf{e}_j - \mathbf{e}_{j-1})\right]\right]$$

$$\overset{(c)}{\geqslant} 0$$

where, (a) is from $W_t^{(n)}\left(\mathbf{x}_t^{(-l)}, \mathbf{x}_t^{(n)}\right) = W_t^{(l)}\left(\mathbf{x}_t^{(-l)}, \mathbf{x}_t^{(n)}\right)$ since the information states of both channel n and channel l are the same value $\mathbf{x}_t^{(n)}$ which implies choosing n or l leads to the same reward, (b) is from Proposition 4.8, and the inequality (c) is from Proposition 4.9, and $\sum_{i=j}^{X}\left(x_t^{(l)}(i) - x_t^{(n)}(i)\right) \geqslant 0$ is due to $\mathbf{x}_t^{(l)} \geqslant_s \mathbf{x}_t^{(n)}$ by Proposition 4.1. ■

Remark 4.4 Lemma 4.3 states that scheduling the channel with better information state would bring more reward.

Based on Lemma 4.3, we have the following theorem which states the optimal condition of the myopic policy.

Theorem 4.1 *Under Assumption* 1, *the myopic policy is optimal.*

Proof When $T \to \infty$, we prove the theorem by backward induction. The theorem holds trivially for T. Assume that it holds for $T-1, \cdots, t+1$, i.e., the optimal accessing policy is to access the best channels from time slot $t+1$ to T. We now show that it holds for t. Suppose, by contradiction, that given $\mathbf{x}_t^{(1)} >_r \cdots >_r \mathbf{x}_t^{(N)}$, the optimal policy is to choose the best from slot $t+1$ to T, and choose $\mu_t = i_1 \neq 1 = \widehat{\mu}_t$ at slot t where $\widehat{\mu}_t$ is to choose the best in the sense of MLR at slot t according to (4.11). Thus, there must exist *in* at slot t such that $\mathbf{x}_t^{(i_n)} >_r \mathbf{x}_t^{(i_1)}$ It then follows from Lemma 4.3 that $W_t^{(i_n)}\left(\mathbf{x}_t^{(1:N)}\right) > W_t^{(i_1)}\left(\mathbf{x}_t^{(1:N)}\right)$, which contradicts with the assumption that the optimal policy is to choose i_1 at slot t. This contradiction completes our proof for a finite T. Obviously, letting $T \rightarrow \infty$, we finish the proof. ■

4.3.5 Discussion

4.3.5.1 Comparison

In [19], the authors considered the problem of scheduling channel with direct or perfect observation, and then their method is based on the information states of all channels in the sense of FOSD order, that is, the critical property is to keep the information states completely ordered or separated in the sense of FOSD order. However, in the case of indirect or imperfect observation, an observation matrix is introduced to replace the identity matrix $\mathbf{E}$ for the direct observation considered in [19]. Hence, the FOSD is not sufficient to characterize the order of information states, and then the MLR order, a kind of stronger stochastic order, is used to describe the order structure of information states. On the other hand, the approach adopted in this paper is totally different from [19] in deriving the optimality of the myopic policy, and consequently, the proposed sufficient conditions cannot degenerate into those of [19] since this approach is more generic and can be extended to the case of heterogeneous multistate channels.

4.3.5.2 Bounds

The bounds in (C1)–(C3) are not enough tight to drop the non-trivial Assumption 1 (v). Actually, we conjecture the optimality of myopic policy is kept even without the Assumption 1(v). However, due to the constraint of the method adopted in this paper, we cannot obtain better bounds to drop the non-trivial Assumption 1(v). Therefore, one of further directions is to obtain the optimality of myopic policy without Assumption 1(v) by some new methods.

4.3.5.3 Case Study

Consider a downlink scheduling system with $N = 4$ channels, one user, and one base station. Each channel has $X = 3$ states, evolving according $\boldsymbol{A}$. The reward vector is $r = (0.050.200.70)^{\top}$, e.g., 0.05 unit reward would be accrued if the chosen channel is in state 1. The discount factor is $\beta = 1$, and the observation matrix $\boldsymbol{B}$ and the initial information states are set as follows:

$$\boldsymbol{A} = \begin{pmatrix} 0.400.200.40 \\ 0.200.240.56 \\ 0.150.250.60 \end{pmatrix}, \boldsymbol{B} = \begin{pmatrix} 0.980.010.01 \\ 0.100.400.50 \\ 0.010.400.59 \end{pmatrix}$$

$$\boldsymbol{x}_0^{(1)} = \begin{pmatrix} 0.40 \\ 0.20 \\ 0.40 \end{pmatrix}, \boldsymbol{x}_0^{(2)} = \boldsymbol{x}_0^{(3)} = \begin{pmatrix} 0.20 \\ 0.24 \\ 0.56 \end{pmatrix}, x_0^{(4)} = \begin{pmatrix} 0.15 \\ 0.25 \\ 0.60 \end{pmatrix}$$

In order to maximize throughput, the base station needs to decide which channel should be used to transmit information to the user each time. Under this setting, it is easy to check that Conditions .1.1–1.5 are satisfied by noticing the following:

$$\phi(\boldsymbol{A}_1^\top, 2) = \begin{pmatrix} 0.0087 \\ 0.3054 \\ 0.6859 \end{pmatrix} \geqslant_r \begin{pmatrix} 0.2000 \\ 0.2400 \\ 0.5600 \end{pmatrix} = \boldsymbol{A}^\top \boldsymbol{A}_3^\top$$

$$\phi(\boldsymbol{A}_3^\top, 1) = \begin{pmatrix} 0.8688 \\ 0.1064 \\ 0.0248 \end{pmatrix} \leqslant_r \begin{pmatrix} 0.2600 \\ 0.2280 \\ 0.5120 \end{pmatrix} = \boldsymbol{A}^\top \boldsymbol{A}_1^\top$$

$$\boldsymbol{r}^\top(\boldsymbol{e}_2 - \boldsymbol{e}_1) - \boldsymbol{r}^\top \boldsymbol{Q}^\top(\boldsymbol{e}_2 - \boldsymbol{e}_1) = 0.0053 \geqslant 0$$

$$\boldsymbol{r}^\top(\boldsymbol{e}_3 - \boldsymbol{e}_2) - \boldsymbol{r}^\top \boldsymbol{Q}^\top(\boldsymbol{e}_3 - \boldsymbol{e}_2) = 0.4638 \geqslant 0$$

Thus, Theorem 4.1 shows that the myopic policy is optimal. That is, in time slot 0, choosing channel 4 is optimal for the base station to communicate with the user since $x_0^{(1)} \leqslant_r x_0^{(2)} \leqslant_r x_0^{(3)} \leqslant_r x_0^{(4)}$ For the other time slots, the optimal policy is to choose the channel with the largest information state, i.e., according to the order of $x_t^{(1)}, x_t^{(2)}, x_t^{(3)}, x_t^{(4)}$.

4.4 Optimality Extension

In this section, we extend the obtained optimality results to the case in which the transition matrix **A** is totally negative ordered in the sense of MLR, as a complementary to the totally positive order discussed in the previous section, which means that those relative propositions are stated here by replacing increasing monotonicity with deceasing monotonicity.

4.4.1 Assumptions

Some important assumptions are stated in the following.

Assumption 2 *Assume that*

(i) $A_1 \geqslant_r A_2 \geqslant_r \cdots \geqslant_r A_X$
(ii) $\mathbf{B}_{:,\,1} \leqslant_r \mathbf{B}_{:,\,2} \leqslant_r \cdots \leqslant_r \mathbf{B}_{:,\,Y}$
(iii) *There exists some* $K(2 \leqslant K \leqslant Y)$ *such that*

$$\Gamma(\mathbf{A}^\top \mathbf{e}_X, K) \leqslant_r (\mathbf{A}^\top)^2 \mathbf{e}_1$$

$$\Gamma(\mathbf{A}^\top \mathbf{e}_1, K-1) \geqslant_r (\mathbf{A}^\top)^2 \mathbf{e}_X$$

(iv) $\mathrm{A}_1 \geqslant_r \mathbf{x}_0^{(1)} \geqslant_r \mathbf{x}_0^{(2)} \geqslant_r \cdots \geqslant_r \mathbf{x}_0^{(N)} \geqslant_r \mathbf{A}_X$
(v) $\mathbf{r}^\top(\mathbf{e}_{i+1} - \mathbf{e}_i) \geqslant \mathbf{r}^\top \mathbf{Q}^\top(\mathbf{e}_{i+1} - \mathbf{e}_i)(1 \leqslant i \leqslant X-1)$, where $\mathbf{A} = \mathbf{V}\Lambda\mathbf{V}^{-1}$, $\mathbf{Q} = \mathbf{V}\Upsilon\mathbf{V}^{-1}$

Remark 4.5 Assumption 2 differs from Assumption 1 in three aspects, i.e., 2.1, 2.3, 2.4, which reflects the inverse TP2 order [4] in matrix **A**.

4.4.2 Optimality

Under Assumption 2, we have the following propositions similar to Propositions 4.3–4.6.

Proposition 4.10 *Let* $\mathbf{x}_1, \mathbf{x}_2 \in \Pi(X)$ *and* $\mathbf{X}_1 \leqslant_r \mathbf{X}_2$*, then*

$$(\mathbf{A}_1)^\top \geqslant_r \mathbf{A}^\top \mathbf{x}_1 \geqslant_r \mathbf{A}^\top \mathbf{x}_2 \geqslant_r (\mathbf{A}_X)^\top$$

Proposition 4.11 Let $\mathbf{x}_1, \mathbf{x}_2 \in \Pi(X)$ and $(\mathbf{A}_1)^\top \geqslant_r \mathbf{x}_1 \geqslant_r \mathbf{x}_2 \geqslant_r (\mathbf{A}_X)^\top$, then for $\Gamma(\mathbf{x}_1, k) \leqslant_r \Gamma(\mathbf{x}_2, k)$ for $1 \leqslant k \leqslant Y$.

Proposition 4.12 Let $\mathbf{X} \in \Pi(X)$ and $(A_1)^\top \geqslant_r x \geqslant_r (A_X)^\top$, then $\Gamma(\mathbf{x}, k) \geqslant_r \Gamma(\mathbf{x}, m)$ for any $1 \leqslant k \leqslant m \leqslant Y$.

Proposition 4.13 Under Assumption 2, we have either $\mathbf{x}_t^{(l)} \leqslant_r \mathbf{x}_t^{(n)}$ or $\mathbf{x}_t^{(n)} \leqslant_r \mathbf{x}_t^{(l)}$ for all $l, n \in \{1, 2, \cdots, N\}$.

Following the similar derivation of Lemma 4.2, we have the following important bounds.

Lemma 4.4 *Under Assumption* 2, $\mathbf{x}_t^l = (\mathbf{x}_t^{(-l)}, \mathbf{x}_t^{(l)}), \check{\mathbf{x}}_t^l = (\mathbf{x}_t^{(-l)}, \check{\mathbf{x}}_t^{(l)}), \mathbf{x}_t^{(l)} \leqslant_r \check{\mathbf{x}}_t^{(l)}$ we have for $1 \leq t \leq T$

(D1) if $u_t' = u_t = l$

$$\mathbf{r}^{\mathrm{T}}\left(\mathbf{E}-\sum_{i=1}^{\lceil\frac{T-t}{2}\rceil}(\beta\mathbf{A}^{\top})^{2i-1}\right)(\check{\mathbf{x}}_t^{(l)}-\mathbf{x}_t^{(l)}) \le W_t^{(u_t')}(\check{\mathbf{x}}_t^{l})-W_t^{(u_t)}(\mathbf{x}_t^{l})$$
$$\le \mathbf{r}^{\top}\left(\mathbf{E}+\sum_{i=1}^{\lfloor\frac{T-t}{2}\rfloor}(\beta\mathbf{A}^{\top})^{2i}\right)(\check{\mathbf{x}}_t^{(l)}-\mathbf{x}_t^{(l)})$$

(D2) if $u_t' \neq l, u_t \neq l$, and $u_t' = u_t$

$$-\mathbf{r}^{\top}\sum_{j=1}^{\lceil\frac{T-t}{2}\rceil}(\beta\mathbf{A}^{\top})^{2i-1}(\check{\mathbf{x}}_t^{(l)}-\mathbf{x}_t^{(l)}) \le W_t^{(u_t')}(\check{\mathbf{x}}_t^{l})-W_t^{(u_t)}(\mathbf{x}_t^{l})$$
$$\le \mathbf{r}^{\top}\sum_{i=1}^{\lfloor\frac{T-t}{2}\rfloor}(\beta\mathbf{A}^{\top})^{2i}(\check{\mathbf{x}}_t^{(l)}-\mathbf{x}_t^{(l)})$$

(D3) if $u_t' = l$ and $u_t \neq l$

$$-\mathbf{r}^{\top}\sum_{i=1}^{\lceil\frac{T-t}{2}\rceil}(\beta\mathrm{A}^{\top})^{2i-1}(\check{\mathbf{x}}_t^{(l)}-\mathbf{x}_t^{(l)}) \le W_t^{(u_t')}(\check{\mathbf{x}}_t^{l})-W_t^{(u_t)}(\mathbf{x}_t^{l})$$
$$\le \mathbf{r}^{\top}\left(\mathbf{E}+\sum_{i=1}^{\lfloor\frac{T-t}{2}\rfloor}(\beta\mathbf{A}^{\top})^{2i}\right)(\check{\mathbf{x}}_t^{(l)}-\mathbf{x}_t^{(l)})$$

Remark 4.6 (D1) achieves its lower bound when l is chosen at slot $t, t+1, t+3, \cdots$, and achieves the upper bound when l is chosen from $t, t+2, t+4, \cdots$. (D2) achieves its lower bound when l is chosen at slot $t+1, t+3, \cdots$, and upper bounds when l is chosen at $t+2, t+4, \cdots$. (D3) achieves its lower bound when l is chosen at slot $t+1$, $t+3, \cdots$, and upper bounds when l is chosen from $t, t+2, t+4, \cdots$.

Based on Lemmas 4.3 and 4.4, we have the following theorem.

Theorem 4.2 *Under Assumption 2, the myopic policy is optimal.*

4.5 Summary

In this chapter, we have investigated the problem of scheduling multistate channels under imperfect state observation. In general, the problem can be formulated as a partially observable Markov decision process or restless multiarmed bandit, which is proved to be PSPACE-hard. In this paper, we have derived a set of closed-form conditions to guarantee the optimality of the myopic policy (scheduling the best channel) in the sense of MLR order. Due to the generic RMAB formulation of the problem, the derived results and the analysis methodology proposed in this paper can be applicable in a wide range of domains.

Appendix

Proof of Lemma 4.1

For slot T, it trivially holds. Suppose it holds for $T-1, \cdots, t+2, t+1$, we prove it holds for slot t.

At slot t, we prove it by two cases in the following.

Case 1: $u_t = n$,

$$
\begin{aligned}
& W_t^{(u_t)}\left(\mathbf{x}_t^{(1:n-1)}, \mathbf{x}_t^{(n)}, \mathbf{x}_t^{(n+1:N)}\right) \\
&= \mathbf{r}^\top \mathbf{x}_t^{(n)} + \beta \sum_{m\in\mathcal{Y}} d\left(\mathbf{x}_t^{(n)}, m\right) W_{t+1}^{\left(\widehat{u_{t+1}}\right)}\left(\mathbf{x}_{t+1}^{(1:n-1)}, \mathbf{x}_{t+1,m}^{(n)}, \mathbf{x}_{t+1}^{(n+1:N)}\right) \\
&\overset{(a)}{=} \mathbf{r}^\top \mathbf{x}_t^{(n)} + \beta \sum_{m\in\mathcal{Y}} d(\mathbf{x}_t^{(n)}, m) \sum_{j=1}^{X} \mathbf{e}_j^\top \mathbf{x}_{t+1,m}^{(n)} W_{t+1}^{\left(\widehat{u_{t+1}}\right)}\left(\mathbf{x}_{t+1}^{(1:n-1)}, \mathbf{e}_j, \mathbf{x}_{t+1}^{(n+1:N)}\right)
\end{aligned}
\tag{4.19}
$$

where the equality (a) is due to the induction hypothesis.

$$
\begin{aligned}
& \sum_{i=1}^{X} x_t^{(n)}(i) W_t^{(u_t)}\left(\mathbf{x}_t^{(1:n-1)}, \mathbf{e}_i, \mathbf{x}_t^{(n+1:N)}\right) \\
&= \sum_{i=1}^{X} x_t^{(n)}(i) \left[\mathbf{r}^\top \mathbf{x}_t^{(n)} + \beta \sum_{m\in\mathcal{Y}} d(\mathbf{e}_i, m) W_{t+1}^{\left(\widehat{u_{t+1}}\right)}\left(\mathbf{x}_{t+1}^{(1:n-1)}, \Gamma(\mathbf{e}_i, m), \mathbf{x}_{t+1}^{(n+1:N)}\right)\right] \\
&\overset{(b)}{=} \mathbf{r}^\top \mathbf{x}_t^{(n)} + \beta \sum_{i=1}^{X} x_t^{(n)}(i) \sum_{m\in\mathcal{Y}} d(\mathbf{e}_i, m) W_{t+1}^{\left(\widehat{u_{t+1}}\right)}\left(\mathbf{x}_{t+1}^{(1:n-1)}, \Gamma(\mathbf{e}_i, m), \mathbf{x}_{t+1}^{(n+1:N)}\right) \\
&\overset{(c)}{=} \mathbf{r}^\top \mathbf{x}_t^{(n)} + \beta \sum_{i=1}^{X} x_t^{(n)}(i) \sum_{m\in\mathcal{Y}} d(\mathbf{e}_i, m) \\
&\quad \times \sum_{j=1}^{X} \mathbf{e}_j^\top \Gamma(\mathbf{e}_i, m) W_{t+1}^{\left(\widehat{u_{t+1}}\right)}\left(\mathbf{x}_{t+1}^{(1:n-1)}, \mathbf{e}_j, \mathbf{x}_{t+1}^{(n+1:N)}\right)
\end{aligned}
\tag{4.20}
$$

where, the equality (b) is from $\sum_{i=1}^{X} x_t^{(n)}(i) = 1$, and equality (c) is due to induction hypothesis.

To prove the lemma, it is sufficient to prove the following equation:

$$\sum_{m\in\mathcal{Y}} d\left(\mathbf{x}_t^{(n)},m\right)\sum_{j=1}^{X} \mathrm{e}_j^\top \mathbf{x}_{t+1,m}^{(n)} = \sum_{i=1}^{X} x_t^{(n)}(i)\sum_{m\in\mathcal{F}} d(\mathbf{e}_i,m)\sum_{j=1}^{X} \mathbf{e}_j^\top \Gamma(\mathbf{e}_i,m) \tag{4.21}$$

Now, we have RHS and LHS of (4.21) as follows:

$$\sum_{m\in\mathcal{Y}} d\left(\mathbf{x}_t^{(n)},m\right)\sum_{j=1}^{X} \mathrm{e}_j^\top \mathbf{x}_{t+1,m}^{(n)} = \sum_{m\in\mathcal{Y}} d\left(\mathbf{x}_t^{(n)},m\right)\sum_{j=1}^{X} \mathrm{e}_j^\top \frac{\mathbf{B}(m)\mathbf{A}^\top \mathbf{x}_t^{(n)}}{d\left(\mathbf{x}_t^{(n)},m\right)}$$

$$= \sum_{m\in\mathcal{Y}}\sum_{j=1}^{X} \mathrm{e}_j^\top \mathbf{B}(m)\mathbf{A}^\top \mathbf{x}_t^{(n)} \tag{4.22}$$

$$= \sum_{i=1}^{X} x_t^{(n)}(i)\sum_{m\in\mathcal{Y}} d(\mathbf{e}_i,m)\sum_{j=1}^{X} \mathbf{e}_j^\top \Gamma(\mathbf{e}_i,m) = \sum_{i=1}^{X} x_t^{(n)}(i)\sum_{m\in\mathcal{Y}} d(\mathbf{e}_i,m)\sum_{j=1}^{X} \mathbf{e}_j^\top \frac{\mathbf{B}(m)\mathbf{A}^\top \mathbf{e}_i}{d(\mathbf{e}_i,m)}$$

$$= \sum_{i=1}^{X} x_t^{(n)}(i)\sum_{m\in\mathcal{Y}}\sum_{j=1}^{X} \mathrm{e}_j^\top \mathbf{B}(m)\mathbf{A}^\top \mathbf{e}_i$$

$$= \sum_{m\in\mathcal{Y}}\sum_{j=1}^{X} \mathrm{e}_j^\top \mathbf{B}(m)\mathbf{A}^\top \sum_{i=1}^{X} x_t^{(n)}(i)\mathbf{e}_i$$

$$= \sum_{m\in\mathcal{Y}}\sum_{j=1}^{X} \mathrm{e}_j^\top \mathbf{B}(m)\mathbf{A}^\top \mathbf{x}_t^{(n)} \tag{4.23}$$

Combining (4.22) and (4.23), we have (4.21), and further, prove the lemma.
Case 2: $u_t \neq n$, without loss of generality, assuming $u_t \geqslant n+1$,

$$W_t^{(u_t)}\left(\mathbf{x}_t^{(1:n-1)},\mathbf{x}_t^{(n)},\mathbf{x}_t^{(n+1:N)}\right)$$

$$= \mathbf{r}^\top \mathbf{x}_t^{(u_t)} + \beta\sum_{m\in\mathcal{Y}} d\left(\mathbf{x}_t^{(u_t)},m\right) W_{t+1}^{(\widehat{u_{t+1}})}\left(\mathbf{x}_{t+1}^{(1:u_t-1)},\mathbf{x}_{t+1,m}^{(u_t)},\mathbf{x}_{t+1}^{(u_t+1:N)}\right)$$

$$\overset{(a)}{=} \mathbf{r}^\top \mathbf{x}_t^{(u_t)} + \beta\sum_{m\in\mathcal{Y}} d\left(\mathbf{x}_t^{(u_t)},m\right)\sum_{i=1}^{X} \mathbf{x}_{t+1}^{(n)}(i)$$

$$\times W_{t+1}^{(\widehat{u_{t+1}})}\left(\mathbf{x}_{t+1}^{(1:n-1)},\mathbf{e}_i,\mathbf{x}_{t+1}^{(n+1:u_t-1)},\mathbf{x}_{t+1,m}^{(u_t)},\mathbf{x}_{t+1}^{(u_t+1:N)}\right), \tag{4.24}$$

where, the equality (a) is due to the induction hypothesis.

$$\sum_{i=1}^{X} x_t^{(n)}(i) W_t^{(u_t)}\left(\mathbf{x}_t^{(1:n-1)}, \mathbf{e}_i, \mathbf{x}_t^{(n+1:N)}\right)$$

$$= \sum_{i=1}^{X} x_t^{(n)}(i)\left[\mathbf{r}^\top \mathbf{x}_t^{(u_t)} + \beta \sum_{m\in\mathcal{Y}} d\left(\mathbf{x}_t^{(u_t)}, m\right) \times W_{t+1}^{(\widehat{u_{t+1}})}\left(\mathbf{x}_{t+1}^{(1:n-1)}, \mathbf{e}_i, \mathbf{x}_{t+1}^{(n+1:u_t-1)}, \mathbf{x}_{t+1,m}^{(u_t)}, \mathbf{x}_{t+1}^{(u_t+1:N)}\right)\right]$$

$$\overset{(b)}{=} \mathbf{r}^\top \mathbf{x}_t^{(u_t)} + \beta \sum_{i=1}^{X} x_t^{(n)}(i) \sum_{m\in\mathcal{Y}} d\left(\mathbf{x}_t^{(u_t)}, m\right)$$
$$\times W_{t+1}^{(\widehat{u_{t+1}})}\left(\mathbf{x}_{t+1}^{(1:n-1)}, \mathbf{e}_i, \mathbf{x}_{t+1}^{(n+1:u_t-1)}, \mathbf{x}_{t+1,m}^{(u_t)}, \mathbf{x}_{t+1}^{(u_t+1:N)}\right) \tag{4.25}$$

where, the equality (b) is from $\sum_{i=1}^{X} x_t^{(n)}(i) = 1$.

Combining (4.24) and (4.25), we prove the lemma.

Proof of Lemma 4.2

We prove the lemma by backward induction.

For slot T, we have

1. For it holds that $W_T^{(u'_T)}(\check{\mathbf{x}}_T^l) - W_T^{(u_T)}(\mathbf{x}_T^l) = \mathbf{r}^\top(\check{\mathbf{x}}_T^{(l)} - \mathbf{x}_T^{(l)})$;
2. For $u'_T \neq l, u_T \neq l$ and $u'_T = u_T$, it holds that $W_T^{(u'_T)}(\check{\mathbf{x}}_T^l) - W_T^{(u_T)}(\mathbf{x}_T^l) = 0$;
3. For $u'_T = l$ and $u_T \neq l$ it exists at least one channel n such that $u'_T = n$ and $\check{\mathbf{x}}_T^{(l)} \geqslant_r \mathbf{x}_T^{(n)} \geqslant_r \mathbf{x}_T^{(l)}$ It then holds that $0 \leqslant W_T^{(u'_T)}(\check{\mathbf{x}}_T^l) - W_T^{(u_T)}(\mathbf{x}_T^l) \leqslant \mathbf{r}^\top(\check{\mathbf{x}}_T^{(l)} - \mathbf{x}_T^{(n)})$.

Therefore, Lemma 4.2 holds for slot T.

Assume that Lemma 4.2 holds for $T-1, \cdots, t+1$, then we prove the lemma for slot t.

We first prove the first case: $u'_t = l, u_t = l$. By developing $\check{\mathbf{x}}_t^{(l)}$ and $\mathbf{x}_t^{(l)}$ according to Lemma 4.1, we have the following:

$$F(\check{\mathbf{x}}_t^l, u'_t) = \sum_{m\in\mathcal{Y}} d\left(\check{\mathbf{x}}_t^{(l)}, m\right) \sum_{j\in X} \mathrm{e}_j^\top \Gamma\left(\check{\mathbf{x}}_t^{(l)}, m\right) W_{t+1}^{(\widetilde{\widehat{u_{t+1}}})}\left(\mathbf{x}_{t+1}^{(-l)}, \mathbf{e}_j\right)$$
$$= \sum_{m\in\mathcal{Y}} \sum_{j\in X} \mathrm{e}_j^\top \mathbf{B}(m) \mathrm{A}^\top \check{\mathbf{x}}_t^{(l)} W_{t+1}^{(\widetilde{\widehat{u_{t+1}}})}\left(\mathbf{x}_{t+1}^{(-l)}, \mathbf{e}_j\right) \tag{4.26}$$

$$F(\mathbf{x}_t^l, u_t) = \sum_{m\in\mathcal{Y}} d\left(\mathbf{x}_t^{(l)}, m\right) \sum_{j\in X} \mathrm{e}_j^\top \Gamma\left(\mathbf{x}_t^{(l)}, m\right) W_{t+1}^{(\widehat{u_{t+1}})}\left(\mathbf{x}_{t+1}^{(-l)}, \mathbf{e}_j\right)$$

$$= \sum_{m\in\mathcal{Y}}\sum_{j\in X} \mathbf{e}_j^\top \mathbf{B}(m)\mathbf{A}^\top \mathbf{x}_t^{(l)} W_{t+1}^{(\widetilde{u_{t+1}})}\left(\mathbf{x}_{t+1}^{(-l)}, \mathbf{e}_j\right) \quad (4.27)$$

Furthermore, we have

$$F(\check{\mathbf{x}}_t^l, u_t') - F(\mathbf{x}_t^l, u_t)$$

$$= \sum_{m\in\mathcal{Y}}\sum_{j\in X}\left[\mathbf{e}_j^\top \mathbf{B}(m)\mathbf{A}^\top \check{\mathbf{x}}_t^{(l)} W_{t+1}^{(\widetilde{u_{t+1}})}\left(\mathbf{x}_{t+1}^{(-l)}, \mathbf{e}_j\right) - \mathbf{e}_j^\top \mathbf{B}(m)\mathbf{A}^\top \mathbf{x}_t^{(l)} W_{t+1}^{(\widehat{u_{t+1}})}\left(\mathbf{x}_{t+1}^{(-l)}, \mathbf{e}_j\right)\right]$$

$$\stackrel{(a)}{=} \sum_{m\in\mathcal{Y}}\sum_{j\in X-\{1\}}\left[\mathbf{e}_j^\top \mathbf{B}(m)\mathbf{A}^\top \left(\check{\mathbf{x}}_t^{(l)} - \mathbf{x}_t^{(l)}\right)\left(W_{t+1}^{(\widetilde{u_{t+1}})}\left(\mathbf{x}_{t+1}^{(-l)}, \mathbf{e}_j\right) - W_{t+1}^{(\widehat{u_{t+1}})}\left(\mathbf{x}_{t+1}^{(-l)}, \mathbf{e}_1\right)\right)\right]$$
$$(4.28)$$

where, the equality (a) is due to $x_t^{(l)}(1) = 1 - \sum_{j\in X^{(l)}-\{1\}} x_t^{(l)}(j)$.

Next, we analyze the term in the bracket, $W_{t+1}^{(\widetilde{u_{t+1}})}\left(\mathbf{x}_{t+1}^{(-l)}, \mathbf{e}_j\right) - W_{t+1}^{(\widehat{u_{t+1}})}\left(\mathbf{x}_{t+1}^{(-l)}, \mathbf{e}_1\right)$ of RHS of (4.28) through three cases:

Case 1: if $\widehat{u}_{t+1}' = l$ and $\widehat{u}_{t+1} = l$, according to the induction hypothesis, we have

$$0 \leqslant W_{t+1}^{(\widetilde{u_{t+1}})}\left(\mathbf{x}_{t+1}^{(-l)}, \mathbf{e}_j\right) - W_{t+1}^{(\widehat{u_{t+1}})}\left(\mathbf{x}_{t+1}^{(-l)}, \mathbf{e}_1\right) \leqslant \sum_{i=0}^{T-t-1} \mathbf{r}^\top (\beta\mathbf{A}^\top)^i (\mathbf{e}_j - \mathbf{e}_1)$$

Case 2: if $\widehat{u}_{t+1}' \neq l, \widehat{u}_{t+1} \neq l$, and $\widehat{u}_{t+1}' = \widehat{u}_{t+1}$, according to the induction hypothesis, we have

$$0 \leqslant W_{t+1}^{(\widetilde{u_{t+1}})}\left(\mathbf{x}_{t+1}^{(-l)}, \mathbf{e}_j\right) - W_{t+1}^{(\widehat{u_{t+1}})}\left(\mathbf{x}_{t+1}^{(-l)}, \mathbf{e}_1\right) \leqslant \sum_{i=1}^{T-t-1} \mathbf{r}^\top (\beta\mathbf{A}^\top)^i (\mathbf{e}_j - \mathbf{e}_1)$$

Case 3: if $\widehat{u}_{t+1}' = l$ and $\widehat{u}_{t+1} \neq l$, according to the induction hypothesis, we have

$$0 \leqslant W_{t+1}^{(\widetilde{u_{t+1}})}\left(\mathbf{x}_{t+1}^{(-l)}, \mathbf{e}_j\right) - W_{t+1}^{(\widehat{u_{t+1}})}(\mathbf{x}_{t+1}^{(-l)}, \mathbf{e}_1) \leqslant \sum_{i=0}^{T} \mathbf{r}^\top (\beta\mathbf{A}^\top)^i (\mathbf{e}_j - \mathbf{e}_1)$$

Combining Cases 1–3, we obtain the bounds of $W_{t+1}^{(\widetilde{u_{t+1}})}\left(\mathbf{x}_{t+1}^{(-l)}, \mathbf{e}_j\right) - W_{t+1}^{(\widehat{u_{t+1}})}\left(\mathbf{x}_{t+1}^{(-l)}, \mathbf{e}_1\right)$ as follows:

$$0 \leqslant W_{t+1}^{(\widetilde{u_{t+1}})}\left(\mathbf{x}_{t+1}^{(-l)}, \mathbf{e}_j\right) - W_{t+1}^{(\widehat{u_{t+1}})}\left(\mathbf{x}_{t+1}^{(-l)}, \mathbf{e}_1\right) \leqslant \sum_{i=0}^{T-t-1} \mathbf{r}^\top (\beta\mathbf{A}^\top)^i (\mathbf{e}_j - \mathbf{e}_1)$$

Therefore, we have

$$W_t^{(u_t)}\left(\check{\mathbf{x}}_t^{(l)}\right) - W_t^{(u_l)}\left(\mathbf{x}_t^l\right)$$

$$= \mathbf{r}^\top\left(\check{\mathbf{x}}_t^{(l)} - \mathbf{x}_t^{(l)}\right) + \beta\Big(F(\check{\mathbf{x}}_t^l, u_t') - F(\mathbf{x}_t^l, u_t)\Big)$$

$$= \mathbf{r}^\top\left(\check{\mathbf{x}}_t^{(l)} - \mathbf{x}_t^{(l)}\right) + \beta\sum_{m\in\mathcal{Y}}\sum_{j\in X-\{1\}}$$

$$\times\left[\mathrm{e}_j^\top\mathbf{B}(m)\mathbf{A}^\top\left(\check{x}_t^{(l)} - \mathbf{x}_t^{(l)}\right)\left(W_{t+1}^{(\widehat{u_{t+1}})}\left(\mathbf{x}_{t+1}^{(-l)}, \mathbf{e}_j\right) - W_{t+1}^{(\widehat{u_{t+1}})}\left(\mathbf{x}_{t+1}^{(-l)}, \mathbf{e}_1\right)\right)\right]$$

$$\leqslant\mathbf{r}^\top\left(\check{\mathbf{x}}_t^{(l)} - \mathbf{x}_t^{(l)}\right) + \beta\sum_{m\in\mathcal{Y}}\sum_{j\in X-\{1\}}$$

$$\times\left[\mathbf{e}_j^\top\mathbf{B}(m)\mathbf{A}^\top\left(\check{x}_t^{(l)} - \mathbf{x}_t^{(l)}\right)\left(\sum_{i=0}^{T-t-1}\mathbf{r}^\top(\beta\mathbf{A}^\top)^i\left(\mathbf{e}_j - \mathbf{e}_1\right)\right)\right]$$

$$\times\left[\mathbf{e}_j^\top\mathbf{B}(m)\mathbf{A}^\top\left(\check{x}_t^{(l)} - \mathbf{x}_t^{(l)}\right)\left(\sum_{i=0}^{T-t-1}\mathbf{r}^\top(\beta\mathbf{A}^\top)^i\left(\mathbf{e}_j - \mathbf{e}_1\right)\right)\right]$$

$$= \sum_{i=0}^{T-t}\mathbf{r}^\top(\beta\mathbf{A}^\top)^i\left(\check{\mathbf{x}}_t^{(l)} - \mathbf{x}_t^{(l)}\right)$$

To the end, we complete the proof of the first part, $u_t' = l$ and $u_t = l$, of Lemma 4.2.

Secondly, **we prove the second case** $u_t' \neq l, u_t \neq l$, **and** $u_t' = u_t$, which implies that in this case, $u_t' = u_t$. Assuming $u_t' = u_t = k$, we have:

$$F(\check{\mathbf{x}}_t^l, u_t')$$

$$= \sum_{m\in\mathcal{Y}} d\left(\mathbf{x}_t^{(k)}, m\right)\sum_{j\in X}\mathrm{e}_j^\top\Gamma(\mathbf{x}_t^{(k)}, m)W_{t+1}^{(u_{t+1})}\left(\mathbf{x}_{t+1}^{(-k,-l)}, \mathbf{e}_j, \mathbf{A}^\top\check{\mathbf{x}}_t^{(l)}\right)$$

$$= \sum_{m\in\mathcal{Y}}\sum_{j\in X}\mathrm{e}_j^\top\mathbf{B}(m)\mathbf{A}^\top\mathbf{x}_t^{(k)}W_{t+1}^{(\widehat{u_{t+1}})}\left(\mathbf{x}_{t+1}^{(-k,-l)}, \mathbf{e}_j, \mathbf{A}^\top\check{\mathbf{x}}_t^{(l)}\right) \tag{4.29}$$

$$F(\mathbf{x}_t^l, u_t)$$

$$= \sum_{m\in\mathcal{Y}} d\left(\mathbf{x}_t^{(k)}, m\right)\sum_{j\in X}\mathrm{e}_j^\top\Gamma(\mathbf{x}_t^{(k)}, m)W_{t+1}^{(\widehat{u_{t+1}})}\left(\mathbf{x}_{t+1}^{(-k,-l)}, \mathbf{e}_j, \mathbf{A}^\top\mathbf{x}_t^{(l)}\right)$$

$$= \sum_{m\in\mathcal{Y}}\sum_{j\in X}\mathrm{e}_j^\top\mathbf{B}(m)\mathbf{A}^\top\mathbf{x}_t^{(k)}W_{t+1}^{(\widehat{u_{t+1}})}\left(\mathbf{x}_{t+1}^{(-k,-l)}, \mathbf{e}_j, \mathbf{A}^\top\mathbf{x}_t^{(l)}\right) \tag{4.30}$$

Thus,

$$F(\check{\mathbf{x}}_t^l, u_t') - F(\mathbf{x}_t^l, u_t)$$
$$= \sum_{m\in\mathcal{Y}}\sum_{j\in\mathcal{X}} \mathbf{e}_j^\top \mathbf{B}(m)\mathbf{A}^\top \mathbf{x}_t^{(k)}$$
$$\times\left[W_{t+1}^{(\widetilde{u_{t+1}})}\left(\mathbf{x}_{t+1}^{(-k,-l)}, \mathbf{e}_j, \mathbf{A}^\top \check{\mathbf{x}}_t^{(l)}\right) - W_{t+1}^{(\widehat{u_{t+1}})}\left(\mathbf{x}_{t+1}^{(-k,-l)}, \mathbf{e}_j, \mathbf{A}^\top \mathbf{x}_t^{(l)}\right)\right] \tag{4.31}$$

For the term in the bracket of RHS of (4.31), if 1 is never chosen for $W_{t+1}^{(\widetilde{u_{t+1}})}\left(\mathbf{x}_{t+1}^{(-k,-l)}, \mathbf{e}_j, \mathbf{A}^\top \check{\mathbf{x}}_t^{(l)}\right)$ and $W_{t+1}^{(\widehat{u_{t+1}})}\left(\mathbf{x}_{t+1}^{(-k,-l)}, \mathbf{e}_j, \mathbf{A}^\top \mathbf{x}_t^{(l)}\right)$ from the slot $t + 1$ to the end of time horizon of interest T. That is to say, $\widehat{u}_\tau' \neq l$ and $\widehat{u}_\tau \neq l$ for $t + 1 \leq \tau \leq T$, and further, we have $W_{t+1}^{(u_{t+1}')}(\mathbf{x}_{t+1}^{(-k,-l)}, \mathbf{e}_j, \mathbf{A}^\top \check{\mathbf{x}}_t^{(l)}) - W_{t+1}^{(\widehat{u_{t+1}})}(\mathbf{x}_{t+1}^{(-k,-l)}, \mathbf{e}_j, \mathbf{A}^\top \mathbf{x}_t^{(l)}) = 0$; otherwise, it exists $t^o(t + 1 \leqslant t^o \leqslant T)$ such that one of the following three cases holds.

Case 1: $u_\tau' \neq l$ and $u_\tau \neq l$ for $t \leq \tau \leq t^0 - 1$ while $u_{t^0}' = l$ and $u_{t^0} = l$;

Case 2: $u_\tau' \neq l$ and $u_\tau \neq l$ for $t \leq \tau \leq t^0 - 1$ while $u_{t^0}' \neq l$ and $u_{t^0} = l$ (Note that this case does not exist since $\mathbf{r}^\top[\mathbf{A}^\top]^{t^0-t}\mathbf{x}_t^{(l)} \geqslant \mathbf{r}^\top[\mathbf{A}^\top]^{t^0-t}\mathbf{x}_t^{(l)}$ according to the transition matrix $\mathbf{A}$);

Case 3: $u_\tau' \neq l$ and $u_\tau \neq l$ for $t \leq \tau \leq t^0 - 1$ while $u_{t^0}' = l$ and $u_{t^0} \neq l$.

For Case 1, according to the hypothesis $\left(u_{t^0}' = l \text{ and } u_{t^0} = l\right)$, we have

$$\beta^{t^0-t}\left(W_{t^0}^{(\widetilde{u_{t^0}})}(\check{x}_{t^0}^l) - W_{t^0}^{(\widehat{u_{t^0}})}(\mathbf{x}_{t^0}^l)\right)$$
$$\leq \beta^{t^0-t}\sum_{i=0}^{T-t^o}(\beta\bar{\lambda})^i\mathbf{r}^\top\left(\check{x}_t^{(l)} - \mathbf{x}_{t^0}^{(l)}\right)$$
$$= \beta^{t^0-t}\sum_{i=0}^{T-t^o}\mathbf{r}^\top(\beta\mathbf{A}^\top)^i[\mathbf{A}^\top]^{t^0-t}\left(\check{\mathbf{x}}_t^{(l)} - \mathbf{X}_t^{(l)}\right)$$
$$\overset{(b)}{\leqslant} \beta\sum_{i=0}^{T-t-1}\mathbf{r}^\top(\beta\mathbf{A}^\top)^i\mathbf{A}^\top\left(\check{\mathbf{x}}_t^{(l)} - \mathbf{x}_t^{(l)}\right)$$

where, the inequality (b) is from $t^0 \geqslant t + 1$.

For Case 3, by the induction hypothesis, we have the similar results with Case 1. Combining the results of the three cases, we obtain the following:

$$W_{t+1}^{(\widetilde{u_{t+1}})}\left(\mathbf{x}_{t+1}^{(-k,-l)}, \mathbf{e}_j, \mathbf{A}^\top \check{\mathbf{x}}_t^{(l)}\right) - W_{t+1}^{(\widehat{u_{t+1}})}\left(\mathbf{x}_{t+1}^{(-k,-l)}, \mathbf{e}_j, \mathbf{A}^\top \mathbf{x}_t^{(l)}\right)$$
$$\leqslant \sum_{i=0}^{T-t-1}\mathbf{r}^\top(\beta\mathbf{A}^\top)^i\mathbf{A}^\top\left(\check{\mathbf{x}}_t^{(l)} - \mathbf{x}_t^{(l)}\right) \tag{4.32}$$

Combining (4.32) and (4.31), we have

$$
\begin{aligned}
& W_t^{(u_t')}(\check{\mathbf{x}}_t^l) - W_t^{(u_t)}(\mathbf{x}_t^l) \\
& = \beta(F(\check{\mathbf{x}}_t^l, u_t') - F(\mathbf{x}_t^l, u_t)) \\
& = \beta \sum_{m \in \mathcal{Y}} \sum_{j \in X} \mathrm{e}_j^\top \mathbf{B}(m) \mathbf{A}^\top \mathbf{x}_t^{(k)}
\end{aligned}
$$

$$
\times \left[W_{t+1}^{(\widetilde{\widehat{u_{t+1}}})}\left(\mathbf{x}_{t+1}^{(-k,-l)}, \mathbf{e}_j, \mathbf{A}^\top \check{\mathbf{x}}_t^{(l)}\right) - W_{t+1}^{(\widehat{u_{t+1}})}\left(\mathbf{x}_{t+1}^{(-k,-l)}, \mathbf{e}_j, \mathbf{A}^\top \mathbf{x}_t^{(l)}\right) \right]
$$

$$
\begin{aligned}
& \leqslant \beta \sum_{m \in \mathcal{Y}} \sum_{j \in X} \mathrm{e}_j^\top \mathbf{B}(m) \mathbf{A}^\top \mathbf{x}_t^{(k)} \sum_{i=0}^{T-t-1} \mathbf{r}^\top (\beta \mathbf{A}^\top)^i \mathbf{A}^\top \left(\check{\mathbf{x}}_t^{(l)} - \mathbf{x}_t^{(l)}\right) \\
& = \sum_{j \in X} \mathrm{e}_j^\top \left[\sum_{m \in \mathcal{Y}} \mathbf{B}(m)\right] \mathbf{A}^\top \mathbf{x}_t^{(k)} \sum_{i=0}^{T-t-1} \mathbf{r}^\top (\beta \mathbf{A}^\top)^{i+1} \left(\check{\mathbf{x}}_t^{(l)} - \mathbf{x}_t^{(l)}\right) \\
& = \sum_{j \in X} \mathrm{e}_j^\top \mathbf{E} \mathbf{A}^\top \mathbf{x}_t^{(k)} \sum_{i=0}^{T-t-1} \mathbf{r}^\top (\beta \mathbf{A}^\top)^{i+1} \left(\check{\mathbf{x}}_t^{(l)} - \mathbf{x}_t^{(l)}\right) \\
& = 1_X^\top A^\top \mathbf{x}_t^{(k)} \sum_{i=0}^{T-t-1} \mathbf{r}^\top (\beta \mathbf{A}^\top)^{i+1} \left(\check{\mathbf{x}}_t^{(l)} - \mathbf{x}_t^{(l)}\right) \\
& = 1_X^\top \mathbf{x}_t^{(k)} \sum_{i=0}^{T-t-1} \mathbf{r}^\top (\beta \mathbf{A}^\top)^{i+1} \left(\check{\mathbf{x}}_t^{(l)} - \mathbf{x}_t^{(l)}\right) \\
& = \sum_{i=0}^{T-t-1} \mathbf{r}^\top (\beta \mathbf{A}^\top)^{i+1} \left(\check{\mathbf{x}}_t^{(l)} - \mathbf{x}_t^{(l)}\right) \\
& = \sum_{i=1}^{T-t} \mathbf{r}^\top (\beta \mathbf{A}^\top)^{i} \left(\check{\mathbf{x}}_t^{(l)} - \mathbf{x}_t^{(l)}\right)
\end{aligned}
$$

which completes the proof of Lemma 4.2 when $u_t' \neq l$ and $u_t \neq l$.

Last, **we prove the third case** $u_t' = l$ and $u_t \neq l$, then it exists at least one process $u_t = n$, and its belief vector denoted as $\mathbf{x}_t^{(n)}$, such that $\check{\mathbf{x}}_t^{(l)} \geqslant_r \mathbf{x}_t^{(n)} \geqslant_r \mathbf{x}_t^{(l)}$ we have

$$
W_t^{(u_t')}(\check{\mathbf{x}}_t^l) - W_t^{(u_t)}(\mathbf{x}_t^l)
$$

$$
= W_t^{(l)}\left(\mathbf{x}_t^{(1)}, \cdots, \mathbf{x}_t^{(l-1)}, \check{\mathbf{x}}_t^{(l)}, \mathbf{x}_t^{(l+1)}, \cdots, \mathbf{x}_t^{(N)}\right) - W_t^{(n)}\left(\mathbf{x}_t^{(1)}, \cdots, \mathbf{x}_t^{(l-1)}, \mathbf{x}_t^{(l)}, \mathbf{x}_t^{(l+1)}, \cdots, \mathbf{x}_t^{(N)}\right)
$$

$$
= \left[W_t^{(l)}\left(\mathbf{x}_t^{(1)}, \cdots, \mathbf{x}_t^{(l-1)}, \check{\mathbf{x}}_t^{(l)}, \mathbf{x}_t^{(l+1)}, \cdots, \mathbf{x}_t^{(N)}\right) - W_t^{(n)}\left(\mathbf{x}_t^{(1)}, \cdots, \mathbf{x}_t^{(l-1)}, \mathbf{x}_t^{(n)}, \mathbf{x}_t^{(l+1)}, \cdots, \mathbf{x}_t^{(N)}\right)\right]
$$

$$
\begin{aligned}
&= \left[W_t^{(l)}\left(\mathbf{x}_t^{(1)},\cdots,\mathbf{x}_t^{(l-1)},\check{\mathbf{x}}_t^{(l)},\mathbf{x}_t^{(l+1)},\cdots,\mathbf{x}_t^{(N)}\right) - W_t^{(l)}\left(\mathbf{x}_t^{(1)},\cdots,\mathbf{x}_t^{(l-1)},\mathbf{x}_t^{(n)},\mathbf{x}_t^{(l+1)},\cdots,\mathbf{x}_t^{(N)}\right)\right] \\
&= \left[W_t^{(l)}\left(\mathbf{x}_t^{(1)},\cdots,\mathbf{x}_t^{(l-1)},\check{\mathbf{x}}_t^{(l)},\mathbf{x}_t^{(l+1)},\cdots,\mathbf{x}_t^{(N)}\right) - W_t^{(l)}\left(\mathbf{x}_t^{(1)},\cdots,\mathbf{x}_t^{(l-1)},\mathbf{x}_t^{(n)},\mathbf{x}_t^{(l+1)},\cdots,\mathbf{x}_t^{(N)}\right)\right] \\
&+ \left[W_t^{(n)}\left(\mathbf{x}_t^{(1)},\cdots,\mathbf{x}_t^{(l-1)},\mathbf{x}_t^{(n)},\mathbf{x}_t^{(l+1)},\cdots,\mathbf{x}_t^{(N)}\right) - W_t^{(n)}\left(\mathbf{x}_t^{(1)},\cdots,\mathbf{x}_t^{(l-1)},\mathbf{x}_t^{(l)},\mathbf{x}_t^{(l+1)},\cdots,\mathbf{x}_t^{(N)}\right)\right]
\end{aligned} \tag{4.33}
$$

According to the induction hypothesis ($l \in \mathcal{A}'$ and $l \in \mathcal{A}$), the first term of the RHS of (4.33) can be bounded as follows:

$$
\begin{aligned}
&W_t^{(l)}\left(\mathbf{x}_t^{(1)}, \cdots, \mathbf{x}_t^{(l-1)}, \check{\mathbf{x}}_t^{(l)}, \mathbf{x}_t^{(l+1)}, \cdots, \mathbf{x}_t^{(N)}\right) \\
&-W_t^{(l)}\left(\mathbf{x}_t^{(1)}, \cdots, \mathbf{x}_t^{(l-1)}, \mathbf{x}_t^{(n)}, \mathbf{x}_t^{(l+1)}, \cdots, \mathbf{x}_t^{(N)}\right) \\
&\leqslant \sum_{i=0}^{T-t} \mathbf{r}^\top (\beta \mathbf{A}^\top)^i \left(\check{\mathbf{x}}_t^{(l)} - \mathbf{x}_t^{(n)}\right)
\end{aligned} \tag{4.34}
$$

Meanwhile, the second term of the RHS of (4.33) is inducted by hypothesis ($l \notin \mathcal{A}'$ and $l \notin \mathcal{A}$):

$$
\begin{aligned}
&W_t^{(n)}\left(\mathbf{x}_t^{(1)}, \cdots, \mathbf{x}_t^{(l-1)}, \mathbf{x}_t^{(n)}, \mathbf{x}_t^{(l+1)}, \cdots, \mathbf{x}_t^{(N)}\right) \\
&- W_t^{(n)}\left(\mathbf{x}_t^{(1)}, \cdots, \mathbf{x}_t^{(l-1)}, \mathbf{x}_t^{(l)}, \mathbf{x}_t^{(l+1)}, \cdots, \mathbf{x}_t^{(N)}\right) \\
&\leqslant \sum_{i=1}^{T-t} \mathbf{r}^\top (\beta \mathbf{A}^\top)^i \left(\mathbf{x}_t^{(n)} - \mathbf{x}_t^{(l)}\right)
\end{aligned} \tag{4.35}
$$

Therefore, we have, combining (4.33), (4.34), and (4.35),

$$
W_t^{(u_t')}(\check{\mathbf{x}}_t^l) - W_t^{(u_t)}(\mathbf{x}_t^l) \leqslant \sum_{i=0}^{T-t} \mathbf{r}^\top (\beta \mathbf{A}^\top)^i (\check{\mathbf{x}}_t^{(l)} - \mathbf{x}_t^{(l)})
$$

Thus, we complete the proof of the third part, $u_t' = l$ and $u_t \neq l$, of Lemma 4.2. To the end, Lemma 4.2 is concluded.

References

1. P. Whittle, Restless bandits: Activity allocation in a changing world. J. Appl. Probab. **24**, 287–298 (1988)

2. J.C. Gittins, K. Glazebrook, R.R. Webber, *Multi-Armed Bandit Allocation Indices* (Blackwell, Oxford, U.K., 2011)
3. C.H. Papadimitriou, J.N. Tsitsiklis, The complexity of optimal queueing network control. Math. Oper. Res. **24**(2), 293–305 (1999)
4. A. Muller, D. Stoyan, *Comparison Methods for Stochastic Models and Risk* (Wiley, New York, 2002)
5. J.C. Gittins, Bandit processes and dynamic allocation indices. J. R. Stat. Soc. Ser. B **41**(2), 148–177 (1979)
6. J.C. Gittins, D.M. Jones, A dynamic allocation index for the sequential design of experiments. Prog. Statist., 241–266 (1974)
7. D. Bertsimas, J. Nino-Mora, Restless bandits, linear programming relaxations, and a primal-dual heuristic. Oper. Res. **48**(1), 80–90 (2000)
8. S. Guha, K. Munagala. Approximation algorithms for partial-information based stochastic control with markovian rewards. in *Proc. IEEE Symposium on Foundations of Computer Science (FOCS)*, Providence, RI, Oct 2007
9. S. Guha, K. Munagala. Approximation algorithms for restless bandit problems. in *Proc. ACM-SIAM Symposium on Discrete Algorithms (SODA)*, New York, Jan 2009
10. S. Ahmad, M. Liu. Multi-Channel Opportunistic Access: A Case of Restless Bandits with Multiple Plays. in *Allerton Conference*, Monticello, Il, Sep–Oct, 2009
11. K. Liu, Q. Zhao, B. Krishnamachari, Dynamic multichannel access with imperfect channel state detection. IEEE Trans. Signal Process. **58**(5), 2795–2807 (2010)
12. F. E. Lapiccirella, K. Liu, Z. Ding: Multi-Channel Opportunistic Access Based on Primary ARQ Messages Overhearing. in *Proc. of ICC*, Kyoto, Jun. 2011
13. S. Ahmand, M. Liu, T. Javidi, Q. Zhao, B. Krishnamachari, Optimality of myopic sensing in multichannel opportunistic access. IEEE Trans. Inf. Theory **55**(9), 4040–4050 (2009)
14. K. Wang, L. Chen, Q. Liu, Opportunistic Spectrum access by exploiting primary user feedbacks in underlay cognitive radio systems: An optimality analysis. IEEE J. Sel. Topics Signal Process. **7**(5), 869–882 (2013)
15. K. Wang, Q. Liu, L. Chen, On optimality of greedy policy for a class of standard reward function of restless multi-armed bandit problem. IET Signal Process. **6**(6), 584–593 (2012)
16. K. Wang, Q. Liu, F.C. Lau, Multichannel opportunistic access by overhearing primary ARQ messages. IEEE Trans. Veh. Technol. **62**(7), 3486–3492 (2013)
17. Q. Zhao, B. Krishnamachari, K. Liu, On myopic sensing for Multi-Channel opportunistic access: Structure, optimality, and performance. IEEE Trans. Wirel. Commun. **7**(3), 54135440 (2008)
18. S. Ahmad, M. Liu, T. Javidi, Q. Zhao, B. Krishnamachari, Optimality of myopic sensing in multi-channel opportunistic access. IEEE Trans. Inf. Theory **55**(9), 4040–4050 (2009)
19. Y. Ouyang, D. Teneketzis, On the optimality of myopic sensing in multi-state channels. IEEE Trans. Inf. Theory **60**, 681–696 (2014)
20. K. Wang: Optimality of Myopic Policy for Restless Multiarmed Bandit with Imperfect Observation. in *Proc. of GlobeCom*, pages 1–6, Washington D. C., USA, Dec 2016
21. K. Wang, L. Chen, J. Yu: On optimality of myopic policy in multi-channel opportunistic access. in *Proc. of ICC*, pages 1–6, Kuala Lumpur, Malaysia, May 2016
22. K. Wang, Q. Liu, F. Li, L. Chen, X. Ma, Myopic policy for opportunistic access in cognitive radio networks by exploiting primary user feedbacks. IET Commun. **9**(7), 1017–1025 (2015)

Chapter 5
Whittle Index Policy for Opportunistic Scheduling: Heterogeneous Multistate Channels

5.1 Introduction

5.1.1 Background

We revisit the following opportunistic scheduling system involving a base station, also referred to as a server, different classes of users with heterogeneous demands, and time-varying multistate Markovian channels. Each channel with different states and classes has a different transmission rate, i.e., the evolution of channels is Markovian and class-dependent. For those users connected to (or entering) the system but not served immediately, their waiting costs increase with time. In such an opportunistic scheduling scenario, a central problem is how to exploit the server's capacity to serve the users. This problem can be formalized to the problem of designing an optimum opportunistic scheduling policy to minimize the average waiting cost.

The above opportunistic scheduling problem is fundamental in many classical and emerging wireless communication systems such as mobile cellular systems including 4G LTE and the emerging 5G, heterogeneous networks (HetNet).

5.1.2 State of The Art

Due to its fundamental importance, the opportunistic scheduling problem has attracted a large body of research on channel-aware schedulers addressing one or more system performance metrics in terms of throughput, fairness, and stability [1–26].

K. Wang, L. Chen, *Restless Multi-Armed Bandit in Opportunistic Scheduling*,
https://doi.org/10.1007/978-3-030-69959-8_5

The seminal work in [2] showed that the system capacity can be improved by opportunistically serving users with maximal transmission rate. Such a scheduler is called cp-rule or *MaxRate* scheduler. In fact, the *MaxRate* scheduler was myopically throughput-optimal, i.e., maximizing the current slot transmission rate but ignoring the impact of the current scheduling on the future throughput, and consequently, was shown to perform badly in system stability from the long-term viewpoint. For instance, the number of waiting users in the system explodes with the increase of system load. Meanwhile, the *MaxRate* scheduler does not fairly schedule those users with lower transmission rates.

To balance system throughput and fairness, the Proportionally Fair (PF) scheduler was proposed and implemented in CDMA 1xEV-DO system of 3G cellular networks [3]. Technically, the PF scheduler maximizes the logarithmic throughput of the system rather than traditional throughput, and as a result provides better fairness [4]. In [5], the authors approximated the PF by the relatively best (RB) scheduler, and analyzed the flow-level stability of the PF scheduler. Actually, the RB scheduler gives priority to users according to their ratio of the current transmission rate to the mean transmission rate. Accordingly, it is fair to users by taking future evolution into account. Consequently, it can provide a minimal throughput to the users with low accessible transmission rates, at the price of being not maximally stable at flow level [6].

Other schedulers, e.g., *scored based (SB)* [7], *proportionally best (PB)*, and *potential improvement (PI)*, belong to the family of the best-condition schedulers. These schedulers give priority to users according to their respective evaluated channel condition and, accordingly, do not have a direct association with transmission rate. They are not myopically throughput-optimal, but rather have a good performance in the long term. They are maximally stable [8, 9], but they do not consider fairness.

The above papers all assume independent and identically distributed (i.i.d.) channels. For the more challenging scenarios, there exist some work on homogeneous channels [10, 11, 18], i.e., i.i.d. in slots, and heterogeneous channels [12–16], i.e., discrete-time Markov process in slots. Under the Markovian channel model, the opportunistic scheduling problem can be mathematically recast to a RMAB [17]. The RMAB is of fundamental importance in stochastic decision theory due to its generic nature, and has application in numerous engineering problems. The central problem in investigating and solving an instance of RMAB is to establish its *indexability*. Once the indexability is established, an index policy can be constructed by assigning an index for each state of each arm to measure the reward of activating the arm at that state. The policy thus consists of simply activating those arms with the largest indices.

In the context of opportunistic scheduling, the authors [18] considered a flow-level scheduling problem with time-homogeneous channel state transition where the probability of being in a state is fixed for any time slot, regardless of the evolution of the system. For the same channel model, the authors [10] considered the opportunistic scheduling problem under the assumption of traffic size following the Pascal distribution. In [10, 11, 18], the indexability was first proved and then the similar closed-form Whittle index was obtained [17]. For heterogeneous channels, the authors of [12–14] considered a generic flow-level scheduling problem with heterogeneous channel state transition, but carried out their work based on a conjecture that the problem is indexable. As a result, they can only verify the indexability of the proposed policy for some specific scenarios by numerical test before computing the policy index. *The indexability of the opportunistic scheduling for the heterogeneous multistate Markovian channels, despite its theoretical and practical importance, is still open today.*

5.1.3 Main Results and Contributions

To bridge the above theoretical gap, we first carry out a deep investigation into the indexability of the heterogeneous channel case formulated in [12–14] and mathematically, identify a set of sufficient conditions on the channel state transition matrix under which the indexability is guaranteed and consequently the Whittle index policy is feasible. Second, by exploiting the structural property of the channel state transition matrix, we obtain the closed-form Whittle index. Third, for a generic channel state transition matrix not satisfying the sufficient conditions, we propose an eigenvalue-arithmetic-mean scheme to approximate this matrix such that the approximate matrix satisfies the sufficient conditions and further the approximate Whittle index is easily obtained. Finally, we present a scheduling algorithm based on the Whittle index, and conduct extensive numerical experiments which demonstrate that the proposed scheduling algorithm can efficiently balance waiting cost and stability.

Our work thus constitutes a small step toward solving the opportunistic scheduling problem in its generic form involving multistate Markovian channels. As a desirable feature, the indexability conditions established in this work only depend on channel state transition matrix without imposing constraints on those user-dependent parameters such as service rate and waiting cost.

Notation $\boldsymbol{e}_i$ denotes an N-dimensional column vector with 1 in the i-th element and 0 in others. $\boldsymbol{I}$ denotes the $N \times N$ identity matrix. 1_N denotes an N-dimensional column vector with 1 in all elements. $\mathbf{0}_N$ denotes an N-dimensional column vector with 0 in all elements. $\mathbf{1}_N^k$ denotes the N-dimensional column vector with 1 in the first k elements and 0 in the remaining $N - k$ elements. Diag $(a_1, \ldots, a_K)$ denotes a diagonal and a block-diagonal matrix with $a_1, \ldots, a_K$. trace(•) denotes the sum of all elements in a diagonal of a matrix. $(\cdot)^\top$ represents the transpose of a matrix or a vector. $(\cdot)^{-1}$ represents the inverse of a matrix.

Table 5.1 Main notations

Symbols	Descriptions
k	User class
$\mathcal{K}$	User class set
t	Slot index
T	Slot set {0, 1, ...}
N_k	The number of channel states of class k user
$\mathcal{N}'_k$	Set {1, 2, ..., N_k}
$Z_k(t)$	Channel state of a class-k user at t
$q_{k,n,m}$	Probability from n to m of class-k user
$s_{k,n}$	Transmission rate of class-k user in state n
$\mu_{k,n}$	Departure probability of class-k user in state n
c_k	Waiting cost of class-k user
$\mathcal{A}_k$	Decision set {0, 1}
$\mathcal{N}_k$	Set {0, 1, ..., N_k}
$\mathbb{W}^a_{k,n}$	The expected one-slot capacity consumption for class k user in channel state n with action a
$\mathbb{R}^a_{k,n}$	The expected one-slot reward for class k user in channel state n with action a
δ_{N-n}	Set {n + 1, ..., N}

5.2 System Model

As mentioned in the introduction, we consider a wireless communication system where a server schedules jobs of heterogeneous users. The system operates in a time-slotted fashion where τ denotes the slot duration and $t \in \mathcal{T}:=\{0, 1, \cdots\}$ denotes the slot index.

Table 5.1 summaries main notations used in this chapter.

5.2.1 *Job, Channel, and User Models*

Suppose that there are K classes of users, $k \in \mathcal{K}:=\{1, 2, \cdots, K\}$. Each user of class k is uniquely associated with a job of class k which is requested from the server and with a dedicated wireless channel of class k through which the job would be transmitted.

Job sizes. The job (or flow) size b_k of users of class k in bits is geometrically distributed with mean $\mathbb{E}\{b_k\} < \infty$ for class $k \in \mathcal{K}$.

Channel condition. For each user, the channel condition varies from slot to slot, independently of all other users. For each class k user, the set of discretized channel conditions is denoted by the finite set $\mathcal{N}'_k:=\{1, 2, \cdots, N_k\}$.

Channel condition evolution. We assume that at each slot, the channel condition of each user in the system evolves according to a class-dependent Markov chain. Thus, for each user of class $k \in \mathcal{K}$, we can define a Markov chain with state space $\mathcal{N}'_k$. We further define $q_{k,n,m} := \mathbb{P}(Z_k(t+1) = m \mid Z_k(t) = n)$, where $Z_k(t)$, denotes the channel condition of a class k user at time t.

The class k channel condition transition probability matrix is thus defined as follows:

$$
\boldsymbol{Q}^{(k)} := \begin{bmatrix} q_{k,1,1} & q_{k,1,2} & \cdots & q_{k,1,N_k} \\ q_{k,2,1} & q_{k,2,2} & \cdots & q_{k,2,N_k} \\ \vdots & \vdots & \ddots & \vdots \\ q_{k,N_k,1} & q_{k,N_k,2} & \cdots & q_{k,N_k,N_k} \end{bmatrix},
$$

where $\sum_{m \in \mathcal{N}'_i} q_{k,n,m} = 1$ for every $n \in \mathcal{N}'_k$.

Transmission rates. When a user of class k is in channel condition $n \in \mathcal{N}'_k$, he can receive data at transmission rate $s_{k,n}$, i.e., his job is served at rate $s_{k,n}$. We assume that the higher the label of the channel condition, the higher the transmission rate, i.e., $0 \leqslant s_{k,1} < s_{k,2} < \cdots < s_{k,N_k}$.

Waiting costs. For every user of class k, the system operator accrues waiting cost $c_k (c_k > 0)$ at the end of every slot if its job is uncompleted.

5.2.2 Server Model

The server is assumed to have *full knowledge* of the system parameters. We investigate the case where the server can serve one user each slot. However, our analysis can be straightforwardly generalized to the case where multiple users can be served each slot. At the beginning of every slot, the server observes the actual channel conditions of all users in the system and decides which user to serve during the slot. We assume that the server is preemptive, i.e., at any time it can interrupt the service of a user whose job is not yet completed. For those jobs not completed, they will be saved and served in the future. The server is also allowed to stay idle, and note that it is not work-conserving because of the time-varying transmission rate. We denote by $\mu_{k,n} :\approx \tau s_{k,n} / \mathbb{E}_0\{\mathrm{b}_k\}$ [12] the departure probability that the job is completed within the current time slot when the server serves a user of class k in channel condition $n \in \mathcal{N}'_k$. Note that the departure probabilities are increasing in the channel condition, i.e., $0 \leqslant \mu_{k,1} < \cdots < \mu_{k,N_k} \leqslant 1$, because the transmission rates satisfy $0 \leqslant s_{k,1} < s_{k,2} < \cdots < s_{k,N_k}$.

5.2.3 Opportunistic Scheduling Problem

In the above opportunistic scheduling model, a central problem is how to maximally exploit the server's capacity to serve users. This problem can be formalized to the problem of designing an optimum opportunistic scheduling policy to minimize the average waiting cost.

5.3 Restless Bandit Formulation and Analysis

In this section we analyze the scheduling problem by the approach of RMAB. For the ease of analysis, we investigate the discounted waiting costs by introducing a discount factor $0 \leqslant \beta < 1$. Basically, the time-average case is a special case where $\beta \to 1$.

5.3.1 Job-Channel-User Bandit

We denote by $\mathcal{A}_k := \{0, 1\}$ the action space of user of class k where action 1 means serving the user and 0 not serving him.

Every job-channel-user couple of class k is characterized by the tuple

$$\left(\mathcal{N}_k, \left(\boldsymbol{w}_k^a\right)_{a \in \mathcal{A}_k}, \left(\boldsymbol{r}_k^a\right)_{a \in \mathcal{A}_k}, \left(\boldsymbol{P}_k^a\right)_{a \in \mathcal{A}_k}\right)$$

where

1. $\mathcal{N}_k := \{0\} \cup \mathcal{N}_k'$ 'is the user state space, where state 0 indicates that the job is completed, and state $n \in \mathcal{N}_k'$ indicates that the current channel condition is n and the job is uncompleted;
2. $\boldsymbol{w}_k^a := \left(w_{k,n}^a\right)_{n \in \mathcal{N}_k}$, where $w_{k,n}^a$ is the expected one-slot capacity consumption, or work required by a user at state n if action a is chosen. Specifically, for every state $n \in \mathcal{N}_k$, $w_{k,n}^1 = 1$ and $w_{k,n}^0 = 0$;
3. $\boldsymbol{r}_k^a := \left(r_{k,n}^a\right)_{n \in \mathcal{N}_k}$, where $r_{k,n}^a$ is the expected one-slot reward earned by a user at state n if action a is selected. Specifically, for every state $n \in \mathcal{N}_k'$, it is the negative of the expected waiting cost, $r_{k,0}^a = 0, r_{k,n}^1 = -\overline{\mu}_{k,n} c_k$ where $\overline{\mu}_{k,n} = 1 - \mu_{k,n}$ and $r_{k,n}^0 = -c_k$.
4. $\boldsymbol{P}_k^a := \left(p_{k,n,m}^a\right)_{n,m \in \mathcal{N}_k}$ where $p_{k,n,m}^a$ is the probability for a user evolving from state n to state m if action a is selected. The one-slot transition probability matrices for action 0 and 1 are as follows:

$$
P_k^0 = \begin{bmatrix} 1 & 0 & \cdots & 0 \\ 0 & q_{k,1,1} & \cdots & q_{k,1,N_k} \\ 0 & q_{k,2,1} & \cdots & q_{k,2,N_k} \\ \vdots & \vdots & \ddots & \vdots \\ 0 & q_{k,N_k,1} & \cdots & q_{k,N_k,N_k} \end{bmatrix}
$$

$$
P_k^1 = \begin{bmatrix} 1 & 0 & \cdots & 0 \\ \mu_{k,1} & \bar{\mu}_{k,1} q_{k,1,1} & \cdots & \bar{\mu}_{k,1} q_{k,1,N_k} \\ \mu_{k,2} & \bar{\mu}_{k,2} q_{k,2,1} & \cdots & \bar{\mu}_{k,2} q_{k,2,N_k} \\ \vdots & \vdots & \ddots & \vdots \\ \mu_{k,N_k} & \bar{\mu}_{k,N_k} q_{k,N_k,1} & \cdots & \bar{\mu}_{k,N_k} q_{k,N_k,N_k} \end{bmatrix}
$$

The dynamics of user j of class k is captured by the state process $\mathrm{x}_k(\cdot)$ and the action process $a_j(\cdot)$, which correspond to state $x_j(t) \in \mathcal{N}_k$ and action $a_j(t) \in \mathcal{A}_k$ at any slot t.

5.3.2 *Restless Bandit Formulation and Opportunistic Scheduling*

Let $\Pi^t_{\mathrm{x,a}}$ denote the set of all the policies composed of actions $a(0)$, $a(1)\cdots$, $a(t)$, where $\mathrm{a}(t)$ is determined by the state history $x(0)$, $x(1)$, $\cdots$, $x(t)$ and the action history $\mathrm{a}(0)$, $\mathrm{a}(1)\cdots$, $\mathrm{a}(t-1)$, i.e.,

$$
\begin{aligned}
\Pi^t_{\mathrm{x,a}} &:= \left\{\mathrm{a}(i) \mid \mathrm{a}(i) = \phi\left(\mathrm{x}^{0:i}, \mathrm{a}^{0:i-1}\right), i = 0, 1\cdots, t\right\} \\
&\overset{(e)}{=} \{\mathrm{a}(i) \mid \mathrm{a}(i) = \phi(\mathrm{x}(i)), i = 0, 1\cdots, t\}
\end{aligned}
$$

where ϕ is a mapping ϕ:$(\mathrm{x}^{0:i}, \mathrm{a}^{0:i-1}) \mapsto \mathrm{a}(i)$, $\mathrm{x}^{0:i} := (\mathrm{x}(0), \cdots, \mathrm{x}(i))$ and $a^{0:i-1} := (a(0), \cdots, a(i-1))$, and (e) is due to the Markovian feature.

Let $\Pi^t_{\mathrm{x,a}}$ denote the space of randomized and non-anticipative policies depending on the joint state process $x := (x_k(\cdot))_{k\in\mathcal{N}}$ and the joint action process $a := (a_k(\cdot))_{k\in X}$, i.e., $\Pi^t_{x,a} = \prod_{k\in X} \Pi^t_{x_k,a_k}$ is the joint policy space.

Let $\mathbb{E}^\pi_\tau$ denote the expectation over the future states $x(\bullet)$ and the action process; $a(\bullet)$, conditioned on past states $x(0)$, $x(1)$, $\cdots$, $x(\tau)$ and the policy $\pi \in \Pi^\tau_{\mathrm{x,a}^\circ}$.

Consider any expected one-slot quantity $G^{\mathrm{a}(t)}_{\mathrm{x}(t)}$ that depends on state $\mathrm{x}(t)$, an action $\mathrm{a}(t)$ at any time slot t. For any policy $\pi \in \Pi^\infty_{\mathrm{x,a}}$ and any discount factor $0 \leqslant \beta < 1$, we define the infinite horizon β-average quantity as follows:

$$\mathbb{B}_0^\pi\{G_{\mathrm{x}(\cdot)}^{\mathrm{a}(\cdot)}, \beta, \infty\} := \lim_{T\to\infty} \frac{\sum_{t=0}^{T-1} \beta^t \mathbb{E}_t^\pi\{G_{\mathrm{x}(t)}^{\mathrm{a}(t)}\}}{\sum_{t=0}^{T-1} \beta^t} \tag{5.1}$$

In the following we consider the discount factor β to be fixed and the horizon to be infinite; therefore, we omit them in $\mathbb{B}_0^\pi\{G_{\mathrm{x}(\cdot)}^{\mathrm{a}(\cdot)}, \beta, \infty\}$ and write briefly $\mathbb{B}_0^\pi\{G_{\mathrm{x}(\cdot)}^{\mathrm{a}(\cdot)}\}$.

The reason for introducing $\mathbb{B}_0^\pi\{\cdot\}$ is that this form can smoothly transit to the average case $\beta = 1$. Henceforth, we always suppose $0 \leqslant \beta < 1$ except when explicitly emphasizing $\beta = 1$.

We are now ready to formulate the opportunistic scheduling problem faced by the server as follows.

Problem 5.1 (Optimum Opportunistic Scheduling) For any discount factor β, the optimum opportunistic scheduling problem is to find a joint policy $\boldsymbol{\pi} = (\pi_1, \cdots, \pi_K) \in \Pi_{\mathrm{x,a}}^\infty$ a maximizing the total discounted reward (i.e., minimizing total discounted cost), mathematically defined as follows.

$$(\mathrm{P}): \max_{\boldsymbol{\pi}\in\Pi_{\mathrm{x,a}}} \mathbb{B}_0^\pi\left\{\sum_{k\in X} r_{k,\mathrm{x}_k(\cdot)}^{\mathrm{a}_k(\cdot)}\right\} \tag{5.2}$$

$$\text{s.t.} \quad \sum_{k\in X} \mathrm{a}_k(t) = 1, t = 0, 1, \cdots \tag{5.3}$$

The constraints (5.3) of problem (P) can be relaxed to the following:

$$\mathbb{E}_t^\pi\left\{\sum_{k\in X} a_k(t)\right\} = 1, t = 0, 1, \cdots$$

$$\Rightarrow \lim_{T\to\infty} \frac{\sum_{t=0}^{T-1} \beta^t \mathbb{E}_t^\pi\left\{\sum_{k\in X} w_{k,\mathrm{x}(t)}^{\mathrm{a}_k(t)}\right\}}{\sum_{t=0}^{T-1} \beta^t} = 1$$

$$\Leftrightarrow \mathbb{B}_0^\pi\left\{\sum_{k\in X} w_{k,x_k}^{a_k(\cdot)}\right\} = 1 \tag{5.4}$$

Using Lagrangian method, we obtain the following by combining (5.2) and (5.4),

$$\max_{\pi\in\Pi_{\mathrm{x,a}}} \mathbb{B}_0^{\pi}\left\{\sum_{k\in X} r_{k,\mathrm{x}_k(\cdot)}^{\mathrm{a}_k(\cdot)}\right\} - \nu\mathbb{B}_0^{\pi}\left\{\sum_{k\in X} w_{k,\mathrm{x}_k(\cdot)}^{\mathrm{a}_k(\cdot)}\right\}$$

$$= \sum_{k\in X}\left(\max_{\pi_k\in\Pi_k,\,2_k} \mathbb{B}_0^{\pi_k}\left\{r_{k,x_k(\cdot)}^{\mathrm{a}_k(\cdot)} - \nu w_{k,x_k(\cdot)}^{\mathrm{a}_k(\cdot)}\right\}\right) \tag{5.5}$$

Thus, we have the subproblem for class $k \in \mathcal{K}$ ::

$$(\text{SP}) : \max_{\pi_k\in I_{x_k,a_k}} \mathbb{B}_0^{\pi_k}\left\{r_{k,x_k(\cdot)}^{a_k(\cdot)} - \nu w_{k,x_k(\cdot)}^{a_k(\cdot)}\right\}. \tag{5.6}$$

Hence, our goal is to find an optimal policy π_k^* for the subproblem $k(k \in \mathcal{K})$ and then construct a feasible joint policy $\pi = \left(\pi_1^*, \cdots, \pi_K^*\right)$ for the problem (P). In the following, we focus on the subproblem (SP) and drop the subscript k.

5.4 Indexability Analysis and Index Computation

In this section, we first give a set of conditions on the channel state transition matrix, and, based on which, we obtain the threshold structure of the optimal scheduling strategy for the subproblem. We then establish the indexability under the proposed conditions.

5.4.1 *Transition Matrices and Threshold Structure*

Condition 1 *Transition matrix $\boldsymbol{Q}$ can be written as*

$$Q = \boldsymbol{O}_0 + \varepsilon_1\boldsymbol{O}_1 + \varepsilon_2\boldsymbol{O}_2 + \cdots + \varepsilon_{2N-2}\boldsymbol{O}_{2N-2}$$

where $\boldsymbol{h} := [h_1, h_2, \cdots, h_N]^\top$, $\boldsymbol{O}_j$ is defined in (5.8) *and ε_j and λ are real numbers satisfying*

$$\lambda_j{:=}\lambda - \varepsilon_{N-j} - \varepsilon_{N-1+j} \leqslant 0, 1\leqslant j < N \tag{5.7}$$

$$\boldsymbol{O}_j{:=}\begin{cases} 1_N(\boldsymbol{h})^\top + \lambda\boldsymbol{I}_N & \text{if } j=0 \\ \underbrace{\left[\boldsymbol{0}_N, \cdots, \boldsymbol{0}_N\right.}_{N-j-1}\left., -\boldsymbol{1}_N^{N-j}, \boldsymbol{1}_N^{N-j}\right] & \text{if } 1\leqslant j\leqslant N-1 \\ \underbrace{\left[0_N, \cdots, 0_N\right.}_{j-N}\left., 1_N - 1_N^{j-N+1}, 1_N^{j-N+1} - 1_N\right] & \text{if } N\leqslant j\leqslant 2N-2 \end{cases} \tag{5.8}$$

Remark 5.1 Basically, Condition 1 implies that

(i) Any two adjacent rows (i.e., $\boldsymbol{Q}i$; $\boldsymbol{Q}i + 1$) of matrix $\boldsymbol{Q}$ differ in only two adjacent positions (i.e., i; $i + 1$). For example, if $N = 3$, $\boldsymbol{Q}$ is written as

$$\boldsymbol{Q} = \begin{bmatrix} h_1 - \varepsilon_2 + \lambda & h_2 - \varepsilon_1 + \varepsilon_2 & h_3 + \varepsilon_1 \\ h_1 + \varepsilon_3 & h_2 - \varepsilon_1 - \varepsilon_3 + \lambda & h_3 + \varepsilon_1 \\ h_1 + \varepsilon_3 & h_2 - \varepsilon_3 + \varepsilon_4 & h_3 - \varepsilon_4 + \lambda \end{bmatrix}$$

(ii) When $\lambda_j = 0$ for all $j(1 \leqslant j < N)$ the $\boldsymbol{Q}$ degenerates into the case of [18].

Now, we give the following lemma on the threshold structure of the optimum scheduling policy for the subproblem.

Lemma 5.1 (Threshold Structure) *Under Condition 1, for every real-valued v, there exists $n \in \mathcal{N} \cup \{-1\}$ such that the optimum scheduling policy only schedules transmission in channel states $\delta_{N-n} := \{m \in \mathcal{N} : m > n\}$.*

Proof Please see Appendix A.1. □

5.4.2 *Indexability Analysis*

For $\pi_k \in \Pi_{x_k, a_k}$, we introduce the concept of serving set, $\delta(\delta \subseteq \mathcal{N}_k)$, such that the user is served if $n \in \delta$ and not served if $n \notin \delta$. By slightly introducing ambiguity, δ can also be regarded as a policy of serving the set δ.

Thus, the subproblem (5.6) can be transformed to

$$\max_{\delta \in \mathcal{N}_k} \mathbb{B}_0^{\delta} \left\{ r_{k, x_k(\cdot)}^{a_k(\cdot)} - \nu w_{k, x_k(\cdot)}^{a_k(\cdot)} \right\} \tag{5.9}$$

For further analysis, we define

$$\mathbb{R}_n^{\delta} := \frac{\mathbb{B}_0^{\delta} \left\{ r_{k, x_k(-)}^{n, \mathbf{a}_k(\cdot)} \right\}}{1 - \beta} \tag{5.10}$$

$$\mathbb{W}_n^{\delta} := \frac{\mathbb{B}_0^{\delta} \left\{ w_{k, x_k(\cdot)}^{n, \mathbf{a}_k(\cdot)} \right\}}{1 - \beta} \tag{5.11}$$

where n refers to the initial state of user of class k.

By Lemma 5.1, if there exists price ν_n for $n \in \mathcal{N}'$ such that both transmitting and not transmitting are optimal for $\nu = \nu_n$, then there exists a set, δ^*, such that both including state n in δ^* and not including state n lead to the same reward, i.e.,

$$\mathbb{R}_n^{\delta^* \cup \{n\}} - \nu_n \mathbb{W}_n^{\delta^* \cup \{n\}} = \mathbb{R}_n^{\delta^* \setminus \{n\}} - \nu_n \mathbb{W}_n^{\delta^* \setminus \{n\}} \tag{5.12}$$

A straightforward consequence is that changing the action only in the initial period must also lead to the same reward, i.e.,

$$\mathbb{R}_n^{\langle 0,\delta^* \rangle} - \nu_n \mathbb{W}_n^{\langle 0,\delta^* \rangle} = \mathbb{R}_n^{\langle 1,\delta^* \rangle} - \nu_n \mathbb{W}_n^{\langle 1,\delta^* \rangle} \tag{5.13}$$

where $\langle a, \delta^* \rangle$ is the policy that employs action a in the initial period and then proceeds according to δ^*.

Then, if $\mathbb{W}_n^{\langle 1,\delta^* \rangle} - \mathbb{W}_n^{\langle 0,\delta^* \rangle} \neq 0$, we have

$$\nu_n = \frac{\mathbb{R}_n^{\langle 1,\delta^* \rangle} - \mathbb{R}_n^{\langle 0,\delta^* \rangle}}{\mathbb{W}_n^{\langle 1,\delta^* \rangle} - \mathbb{W}_n^{\langle 0,\delta^* \rangle}}. \tag{5.14}$$

We further define

$$\nu_n^{\delta} := \frac{\mathbb{R}_n^{\langle 1,\delta \rangle} - \mathbb{R}_n^{\langle 0,\delta \rangle}}{\mathbb{W}_n^{\langle 1,\delta \rangle} - \mathbb{W}_n^{\langle 0,\delta \rangle}} \tag{5.15}$$

To circumvent the long proof of Whittle indexability, we establish the indexability result by checking the LP-indexability condition [27]. If a problem is LP-indexable, then it is Whittle-indexable. In the following analysis, we show that our problem is LP-indexable; that is, the problem is Whittle-indexable.

Definition 5.1 ([27]) Problem (5.6) is LP-indexable with price

$$\nu_n = \nu_n^{\delta_{N-n}} = \frac{\mathbb{R}_n^{\langle 1,\delta_{N-n} \rangle} - \mathbb{R}_n^{\langle 0,\delta_{N-n} \rangle}}{\mathbb{W}_n^{\langle 1,\delta_{N-n} \rangle} - \mathbb{W}_n^{\langle 0,\delta_{N-n} \rangle}}, \tag{5.16}$$

if the following conditions hold:

(i) $n \in \mathcal{N}, \mathbb{W}_n^{\langle 1,0 \rangle} - \mathbb{W}_n^{\langle 0,0 \rangle} > 0, \mathbb{W}_n^{\langle 1,\mathcal{N} \rangle} - \mathbb{W}_n^{\langle 0,\mathcal{N} \rangle} > 0$;
(ii) $n \in \mathcal{N} \setminus \{N\}, \mathbb{W}_n^{\langle 1,\delta_{N-n} \rangle} - \mathbb{W}_n^{\langle 0,\delta_{N-n} \rangle} > 0$ and $\mathbb{W}_{n+1}^{\langle 1,\delta_{N-n} \rangle} - \mathbb{W}_{n+1}^{\langle 0,\delta_{N-n} \rangle} > 0$;
(iii) For each real value ν, there exists $n \in \mathcal{N} \cup \{-1\}$ such that the serving set δ_{N-n} is optimal.

To check the LP-indexability, we first characterize the four critical quantities in (5.16) under δ_{N-n} for any n.

Based on balance equations, when n is not chosen in the initial slot, we have (5.17) in the matrix language (see the top of the next page) and, further, the following simplified form:

$$\begin{bmatrix} \mathbb{R}_1^{\langle 0,\delta_{N-n}\rangle} \\ \vdots \\ \mathbb{R}_n^{\langle 0,\delta_{N-n}\rangle} \\ \mathbb{R}_{n+1}^{\langle 1,\delta_{N-n}\rangle} \\ \vdots \\ \mathbb{R}_N^{\langle 1,\delta_{N-n}\rangle} \end{bmatrix} = -\beta \begin{bmatrix} q_{1,1} & \cdots & q_{1,n-1} & \cdots & q_{1,N} \\ \vdots & \ddots & \vdots & \ddots & \vdots \\ q_{n,1} & \cdots & q_{n,n} & \cdots & q_{n,N} \\ \bar{\mu}_{n+1} q_{n+1,1} & \cdots & \bar{\mu}_{n+1} q_{n+1,n} & \cdots & \bar{\mu}_{n+1} q_{n+1,N} \\ \vdots & \ddots & \vdots & \ddots & \vdots \\ \bar{\mu}_N q_{N,1} & \cdots & \bar{\mu}_N q_{N,n-1} & \cdots & \bar{\mu}_N q_{N,N} \end{bmatrix} \begin{bmatrix} \mathbb{R}_1^{\langle 0,\delta_{N-n}\rangle} \\ \vdots \\ \mathbb{R}_n^{\langle 0,\delta_{N-n}\rangle} \\ \mathbb{R}_{n+1}^{\langle 1,\delta_{N-n}\rangle} \\ \vdots \\ \mathbb{R}_N^{\langle 1,\delta_{N-n}\rangle} \end{bmatrix} + \begin{bmatrix} -c \\ \vdots \\ -c \\ -c\bar{\mu}_{n+1} \\ \vdots \\ -c\bar{\mu}_N \end{bmatrix} \tag{5.17}$$

$$\begin{bmatrix} \mathbb{R}_1^{\langle 0,\delta_{N-n}\rangle} \\ \vdots \\ \mathbb{R}_{n-1}^{\langle 0,\delta_{N-n}\rangle} \\ \mathbb{R}_{n}^{\langle 1,\delta_{N-n}\rangle} \\ \vdots \\ \mathbb{R}_N^{\langle 1,\delta_{N-n}\rangle} \end{bmatrix} = - \begin{bmatrix} q_{1,1} & \cdots & q_{1,n-1} & q_{1,n} & \cdots & q_{1,N} \\ \vdots & \ddots & \vdots & \vdots & \ddots & \vdots \\ q_{n-1,1} & \cdots & q_{n-1,n-1} & q_{n-1,n} & \cdots & q_{n-1,N} \\ \bar{\mu}_n q_{n,1} & \cdots & \bar{\mu}_n q_{n,n-1} & \bar{\mu}_n q_{n,n} & \cdots & \bar{\mu}_n q_{n,N} \\ \vdots & \ddots & \vdots & \vdots & \ddots & \\ \bar{\mu}_N q_{N,1} & \cdots & \bar{\mu}_N q_{N,n-1} & \bar{\mu}_N q_{N,n} & \cdots & \bar{\mu}_N q_{N,N} \end{bmatrix} \begin{bmatrix} \mathbb{R}_1^{\langle 0,\delta_{N-n}\rangle} \\ \vdots \\ \mathbb{R}_{n-1}^{\langle 0,\delta_{N-n}\rangle} \\ \mathbb{R}_{n}^{\langle 1,\delta_{N-n}\rangle} \\ \vdots \\ \mathbb{R}_N^{\langle 1,\delta_{N-n}\rangle} \end{bmatrix} + \begin{bmatrix} -c \\ \vdots \\ -c \\ -c\bar{\mu}_{n} \\ \vdots \\ -c\bar{\mu}_N \end{bmatrix} \tag{5.18}$$

$$(\boldsymbol{I}_N - \beta \boldsymbol{M}_0) \cdot \boldsymbol{r}_0 = \boldsymbol{c}_0 \tag{5.19}$$

where,

$$\boldsymbol{M}_0 = \left[\boldsymbol{Q}_1^\top, \cdots, \boldsymbol{Q}_n^\top, \quad \boldsymbol{Q}_{n+1}^\top \bar{\mu}_{n+1}, \cdots, \boldsymbol{Q}_N^\top \bar{\mu}_N\right]^\top$$

$$\boldsymbol{c}_0 = \left[-c, \cdots, -c, -c, -c\bar{\mu}_{n+1}, \cdots, -c\bar{\mu}_N\right]^\top$$

$$\boldsymbol{r}_0 = \left[\mathbb{R}_1^{\langle 0,\delta_{N-n}\rangle}, \cdots, \mathbb{R}_n^{\langle 0,\delta_{N-n}\rangle}, \mathbb{R}_{n+1}^{\langle 1,\delta_{N-n}\rangle}, \cdots, \mathbb{R}_N^{\langle 1,\delta_{N-n}\rangle}\right]^\top.$$

Similarly, when n is chosen in the initial slot, we have (5.18) and, further, the following:

$$(\boldsymbol{I}_N - \beta \boldsymbol{M}_1) \cdot \boldsymbol{r}_1 = \boldsymbol{c}_1 \tag{5.20}$$

where,

$$\boldsymbol{M}_1 = \left[\boldsymbol{Q}_1^\top, \cdots, \boldsymbol{Q}_{n-1}^\top, \boldsymbol{Q}_n^\top \overline{\mu}_n, \cdots, \boldsymbol{Q}_N^\top \overline{\mu}_N\right]^\top$$
$$\boldsymbol{c}_1 = \left[-c, \cdots, -c, -c\overline{\mu}_n, -c\overline{\mu}_{n+1}, \cdots, -c\overline{\mu}_N\right]^\top$$
$$\boldsymbol{r}_1 = \left[\mathbb{R}_1^{\langle 0,\delta_{N-n}\rangle}, \cdots, \mathbb{R}_{n-1}^{\langle 0,\delta_{N-n}\rangle}, \mathbb{R}_n^{\langle 1,\delta_{N-n}\rangle}, \cdots, \mathbb{R}_N^{\langle 1,\delta_{N-n}\rangle}\right]^\top$$

Thus, from (5.19) and (5.20), we can obtain

$$\mathbb{R}_n^{\langle 0,\delta_{N-n}\rangle} = \boldsymbol{e}_n^\top (\boldsymbol{I}_N - \beta \boldsymbol{M}_0)^{-1} \boldsymbol{c}_0 \tag{5.21}$$
$$\mathbb{R}_n^{\langle 1,\delta_{N-n}\rangle} = \boldsymbol{e}_n^\top (\boldsymbol{I}_N - \beta \boldsymbol{M}_1)^{-1} \boldsymbol{c}_1 \tag{5.22}$$

Similarly, replacing $\boldsymbol{c}_0, \boldsymbol{c}_1$ by $1_N - 1_N^n, 1_N - 1_N^{n-1}$ from (5.19) and (5.20), respectively, we have

$$(\boldsymbol{I}_N - \beta \boldsymbol{M}_0) \cdot \boldsymbol{w}_0 = 1_N - 1_N^n \tag{5.23}$$
$$(\boldsymbol{I}_N - \beta \boldsymbol{M}_1) \cdot \boldsymbol{w}_1 = \mathbf{1}_N - \mathbf{1}_N^{n-1} \tag{5.24}$$

where,

$$\boldsymbol{w}_0 = \left[\mathbb{W}_1^{\langle 0,\delta_{N-n}\rangle}, \cdots, \mathbb{W}_n^{\langle 0,\delta_{N-n}\rangle}, \mathbb{W}_{n+1}^{\langle 1,\delta_{N-n}\rangle}, \cdots, \mathbb{W}_N^{\langle 1,\delta_{N-n}\rangle}\right]^\top$$
$$\boldsymbol{w}_1 = \left[\mathbb{W}_1^{\langle 0,\delta_{N-n}\rangle}, \cdots, \mathbb{W}_{n-1}^{\langle 0,\delta_{N-n}\rangle}, \mathbb{W}_n^{\langle 1,\delta_{N-n}\rangle}, \cdots, \mathbb{W}_N^{\langle 1,\delta_{N-n}\rangle}\right]^\top$$

Further,

$$\mathbb{W}_n^{\langle 0,\delta_{N-n}\rangle} = \boldsymbol{e}_n^\top (\boldsymbol{I}_N - \beta \boldsymbol{M}_0)^{-1} (\mathbf{1}_N - \mathbf{1}_N^n) \tag{5.25}$$
$$\mathbb{W}_n^{\langle 1,\delta_{N-n}\rangle} = \boldsymbol{e}_n^\top (\boldsymbol{I}_N - \beta \boldsymbol{M}_1)^{-1} (1_N - 1_N^{n-1}) \tag{5.26}$$

After obtaining the four critical quantities, we now check the LP-indexability condition.

Lemma 5.2 *Under Condition 1, for any $n \in \mathcal{N} \backslash \{N\}$, we have*

(i) $\mathbb{W}_n^{\langle 1,\delta_{N-n}\rangle} > \mathbb{W}_n^{\langle 0,\delta_{N-n}\rangle}$
(ii) $\mathbb{W}_{n+1}^{\langle 1,\delta_{N-n}\rangle} > \mathbb{W}_{n+1}^{\langle 0,\delta_{N-n}\rangle}$

Proof Please see Appendix A.2. □

Lemma 5.3 *Under Condition 1, Problem* (5.6) *is LP-indexable with price v_n in* (5.16).

Proof According to Definition 5.1, we prove the indexability by checking three conditions.

(i) Obviously, $\mathbb{W}_n^{\langle 0,\varnothing\rangle} = 0, \mathbb{W}_n^{\langle 1,\varnothing\rangle} \geqslant 1$, and $\mathbb{W}_n^{\langle 1,\mathcal{N}\rangle} = \frac{1}{1-\beta}$. For any, $\delta, \mathbb{W}_n^{\delta} \leqslant \frac{1}{1-\beta}$ and further $\mathbb{W}_n^{\langle 0,\mathcal{N}\rangle} < \frac{1}{1-\beta}$.
(ii) The second condition is proved in Lemma 5.2.
(iii) The third condition is proved in Lemma 5.1.

Therefore, the LP-indexability is proved.

Following Lemma 5.3, the following theorem states our main result on the indexability of Problem (5.6). □

Theorem 5.1 (Indexability) ***Under Condition 1, we have***
(i) if $v \leqslant v_n$, it is optimal to serve the user in state n;
(ii) if $v > v_n$, it is optimal not to serve the user in state n.

5.4.3 Computing Index

In this part, we exploit the structural property of the transition matrix $\boldsymbol{Q}$ to simplify the index computation and further obtain the closed-form Whittle index.

Proposition 5.1 ***Under Condition 1, we have***
(i) $\mathbb{W}_1^{\langle 0,\delta_{N-n}\rangle} = \cdots = \mathbb{W}_w^{\langle 0,\delta_{N-n}\rangle} = \mathbb{W}_n^{\langle 0,\delta_{N-n}\rangle}$.
(ii) $\mathbb{R}_1^{\langle 0,\delta_{N-n}\rangle} = \cdots = \mathbb{R}_{n-1}^{\langle 0,\delta_{N-n}\rangle} = \mathbb{R}_n^{\langle 0,\delta_{N-n}\rangle}$
(iii) *The Whittle index is*

$$\nu_n = \frac{-\mu_n \mathbb{R}_n^{\langle 1,\delta_{N-n}\rangle}}{1-\mu_n \mathbb{W}_n^{\langle 1,\delta_{N-n}\rangle}} \tag{5.27}$$

Proof Following the proof of Lemma 5.2, we have

$$\mathbb{W}_1^{\langle 0,\delta_{N-n}\rangle} = \cdots = \mathbb{W}_{n-1}^{\langle 0,\delta_{N-n}\rangle} = \mathbb{W}_n^{\langle 0,\delta_{N-n}\rangle}$$

and

$$\mathbb{W}_n^{\langle 1,\delta_{N-n}\rangle} - \mathbb{W}_n^{\langle 0,\delta_{N-n}\rangle} = \frac{1-\mu_n \mathbb{W}_n^{\langle 1,\delta_{N-n}\rangle}}{1-\mu_n}\left[\boldsymbol{e}_n^\top (\boldsymbol{I}_N - \beta \boldsymbol{M}_0)^{-1}\boldsymbol{e}_n\right] \tag{5.28}$$

Similarly,

$$\mathbb{R}_1^{\langle 0,\delta_{N-n}\rangle} = \cdots = \mathbb{R}_{n-1}^{\langle 0,\delta_{N-n}\rangle} = \mathbb{R}_n^{\langle 0,\delta_{N-n}\rangle}$$

and

$$\mathbb{R}_n^{\langle 1,\delta_{N-n}\rangle} - \mathbb{R}_n^{\langle 0,\delta_{N-n}\rangle} = \frac{-\mu_n \mathbb{R}_n^{\langle 1,\delta_{N-n}\rangle}}{1-\mu_n}\left[\boldsymbol{e}_n^\top (\boldsymbol{I}_N - \beta \boldsymbol{M}_0)^{-1}\boldsymbol{e}_n\right] \tag{5.29}$$

Therefore,

$$\nu_n = \frac{\mathbb{R}_n^{\langle 1,\delta_{N-n}\rangle} - \mathbb{R}_n^{\langle 0,\delta_{N-n}\rangle}}{\mathbb{W}_n^{\langle 1,\delta_{N-n}\rangle} - \mathbb{W}_n^{\langle 0,\delta_{N-n}\rangle}} = \frac{-\mu_n \mathbb{R}_n^{\langle 1,\delta_{N-n}\rangle}}{1-\mu_n \mathbb{W}_n^{\langle 1,\delta_{N-n}\rangle}}$$

Based on (5.27), in order to obtain ν_n, we only need to compute $\mathbb{W}_n^{\langle 1,\delta_{N-n}\rangle}$ and $\mathbb{R}_n^{\langle 1,\delta_{N-n}\rangle}$. Further, by some complex operations, we can obtain the closed-form Whittle index as follows:

$$\nu_n = \frac{c\mu_n}{1-\beta+f(n)}, 1 \leqslant n \leqslant N \tag{5.30}$$

where

$$f(n) = \sum_{i=n+1}^{N}\left(\beta q_{n,i}\left(1-\prod_{j=n+1}^{i}\frac{\frac{1}{\overline{\mu}_{j-1}}-\beta\lambda_{j-1}}{\frac{1}{\overline{\mu}_j}-\beta\lambda_{j-1}}\right)+\mu_n d_{n-1,i}K_i\right)$$

$$d_{n-1,i} = -\beta\left(q_{n,i}+\sum_{k=i+1}^{N} q_{n,k}\prod_{j=i+1}^{k}\frac{\frac{1}{\overline{\mu}_{j-1}}-\beta\lambda_{j-1}}{\frac{1}{\overline{\mu}_j}-\beta\lambda_{j-1}}\right)$$

$$K_i = \frac{\frac{1}{1-\mu_i}-\frac{1}{1-\mu_{i-1}}}{\frac{1}{1-\mu_i}-\beta\lambda_{i-1}}$$

for $i(n+1 \leq i \leq N)$. □

5.5 Indexability Extension and Scheduling Policy

In this section, we first extend the proposed Condition 1 and obtain the indexability as well as the Whittle index. Next, we propose an eigenvalue-arithmetic-mean scheme to approximate any transition matrix, and further obtain the corresponding approximate Whittle index. Finally, based on the closed-form Whittle index, we construct an efficient scheduling policy.

5.5.1 *Indexability Extension*

In Sect. 5.4.3, the computing process of V_n shows that the V_n only depends on the structure of Q rather than the sign of λ_j, i.e., (5.7). Thus, we release Condition 1 based on the monotonicity of vn and obtain the following theorem on the indexability.

Theorem 5.2 *If* $\boldsymbol{Q}$ *can be written as*

$$\boldsymbol{Q} = \boldsymbol{O}_0 + \varepsilon_1 \boldsymbol{O}_1 + \varepsilon_2 \boldsymbol{O}_2 + \cdots + \varepsilon_{2N-2} \boldsymbol{O}_{2N-2}, \tag{5.31}$$

then Problem (5.6) *is indexable and the Whittle index for state* $n(n = 1, \cdots, N)$ *is*

$$v_n = \begin{cases} \infty, \text{if } \beta = 1, n = N, \\ \dfrac{c\mu_n}{1-\beta+f(n)}, \text{otherwise}. \end{cases} \tag{5.32}$$

Proof Please see the Appendix A.3. □

Remark 5.2 The critical constraint (5.7) of Condition 1 is deleted in this theorem compared with Lemma 5.3.

Corollary 5.1 *If* $\boldsymbol{Q}$ *can be written as*

$$\boldsymbol{Q} = \boldsymbol{O}_0 + \varepsilon_1 \boldsymbol{O}_1 + \varepsilon_2 \boldsymbol{O}_2 + \cdots + \varepsilon_{2N-2} \boldsymbol{O}_{2N-2}$$

and $\lambda_1 = \cdots = \lambda_{N-1} = \lambda$, *then Problem* (5.6) *is indexable and the Whittle index for state* $n(n = 1, \cdots, N)$ *is*

$$v_n = \begin{cases} \infty, & \text{if } \beta = 1, n = N, \\ \dfrac{c\mu_n}{1-\beta+\beta \sum\limits_{i=n+1}^{N} \dfrac{q_{n,i}(\mu_i - \mu_n)}{1-\beta\lambda(1-\mu_i)}}, & \text{otherwise}. \end{cases} \tag{5.33}$$

Remark 5.3 This corollary shows that the Whittle index degenerates into that of [18] if $\lambda_1 = \cdots = \lambda_{N-1} = \lambda = 0$.

5.5.2 Transition Matrix Approximation

Given a generic $\boldsymbol{Q}$, where

$$\boldsymbol{Q} \neq \boldsymbol{O}_0 + \varepsilon_1 \boldsymbol{O}_1 + \varepsilon_2 \boldsymbol{O}_2 + \cdots + \varepsilon_{2N-2} \boldsymbol{O}_{2N-2},$$

thus the result of Theorem 5.2 cannot be used.

For this case, we approximate $\boldsymbol{Q}$ by the following eigenvalue-arithmetic-mean scheme,

$$\boldsymbol{Q} = \boldsymbol{V} \boldsymbol{\Lambda} \boldsymbol{V}^{-1} \tag{5.34}$$

$$\widehat{\boldsymbol{Q}} = \boldsymbol{V} \widehat{\boldsymbol{\Lambda}} \boldsymbol{V}^{-1} \tag{5.35}$$

$$\widehat{\boldsymbol{\Lambda}} = \mathrm{diag}\left(1, \widehat{\lambda}, \cdots, \widehat{\lambda}\right) \tag{5.36}$$

$$\widehat{\lambda} = \frac{\mathrm{trace}(\boldsymbol{Q}) - 1}{N - 1} \tag{5.37}$$

where $\widehat{\lambda}$ is the arithmetic mean of the $N-1$ eigenvalues of Q (excluding the trivial eigenvalue 1). Thus, the approximate matrix $\boldsymbol{Q}$ satisfies the condition of Corollary 5.1, and furthermore, the Whittle index can be approximated by

$$V_n = \begin{cases} \infty, & \text{if } \beta = 1, n = N, \\ \dfrac{c\mu_n}{1 - \beta + \beta \sum\limits_{i=n+1}^{N} \dfrac{q_{n,i}(\mu_i - \mu_n)}{1 - \beta\widehat{\lambda}(1 - \mu_i)}}, & \text{otherwise.} \end{cases} \tag{5.38}$$

5.5.3 Scheduling Policy

In the previous sections, we have obtained the closed-form Whittle index for each subproblem. Now, we construct the joint scheduling policy for the original problem.

In particular, the scheduling policy is to serve the user in $k^*(t)$ with the highest actual price, i.e.,

$$k^*(t) = \text{argmax}_{k \in \mathcal{K}}[\nu_{k,x_k(t)}], \text{if } \nu_{k,x_k(t)} < \infty. \tag{5.39}$$

Actually, $\nu_{k,x_k(t)} < \infty$ always holds if $0 \leqslant \beta < 1$. It happens $\nu_{k,x_k(t)} \to \infty$ only when $\beta = 1$ and $x_k(t) = N_k$, corresponding to the average case.

Therefore, the second item, $c_k\mu_{k,x_k(t)}$, of Laurent expansion of $\nu_{k,x_k(t)}$ would be taken as the secondary index in the case of $\beta = 1$ and $x_k(t) = N_k$ since

$$\lim_{\beta \to 1} (1-\beta)\nu_{k,N_k} = \frac{(1-\beta)c_k\mu_{k,N_k}}{1-\beta} = c_k\mu_{k,N_k}. \tag{5.40}$$

Now, we give the marginal productivity index (MPI) scheduler in Algorithm 1. The MPI scheduler always serves the user currently with the best condition, i.e., $v_1 \leqslant \cdots \leqslant v_N$, and is one of the best-condition schedulers, which has the stability property in a Markovian setting [9].

Theorem 5.3 ([9]) *The MPI scheduler with one server is maximally stable under arbitrary arrivals.*

Algorithm 1 MPI scheduler ($\beta = 1$)

for $t \in \mathcal{T}$
  $C \leftarrow$ number of system users in $N_k (k \in \mathcal{K})$
if $C \geq 1$ **then**
  Serve one user in N_k with $\max\{c_k\mu_{k,N_k}\}(k \in \mathcal{K})$
  (breaking ties randomly)
else
  if condition (5.31) is satisfied
    Serve the user $k^*(t)$ with highest index value by (5.33)
  else
    Serve the user $k^*(t)$ with highest index value by (5.38)
  end if
    (breaking ties randomly)
end if
end for

5.6 Numerical Simulation

In this section, we compare the proposed MPI scheduler with the following policies:

(i) the $c\mu$ rule, $\nu_{k,n}^{c\mu} = c_k\mu_{k,n}$,
(ii) the RB rule, $\nu_{k,n}^{\text{RB}} = \frac{c_k\mu_{k,n}}{\sum_{m=1}^{N_k} q_{k,m}^{\text{SS}}\mu_{k,m}}$,

Table 5.2 Parameters adopted in simulation

No.	$S_{k,n}$ (Mb/s)	(c_1, c_2)	Channel Transition Matrices	Job size (MB)
1	8:4 50:4 53:76 26:88 44:688 80:64	(1, 1)	$\begin{bmatrix} 0.00 & 0.80 & 0.20 \\ 0.30 & 0.50 & 0.20 \\ 0.30 & 0.60 & 0.10 \end{bmatrix}, \begin{bmatrix} 0.00 & 0.80 & 0.20 \\ 0.30 & 0.50 & 0.20 \\ 0.30 & 0.60 & 0.10 \end{bmatrix}$	0.5, 0.5
2	8:4 16:8 33:6 8:4 16:8 33:6	(10, 1)	$\begin{bmatrix} 0.00 & 0.50 & 0.50 \\ 0.10 & 0.40 & 0.50 \\ 0.10 & 0.70 & 0.20 \end{bmatrix}, \begin{bmatrix} 0.25 & 0.60 & 0.15 \\ 0.35 & 0.50 & 0.15 \\ 0.35 & 0.55 & 0.10 \end{bmatrix}$	5, 0.5
3	8:4 16:8 50:4 67:2 26:88 33:6 44:688 80:64	(2,3)	$\begin{bmatrix} 0.50 & 0.10 & 0.20 & 0.20 \\ 0.15 & 0.45 & 0.20 & 0.20 \\ 0.15 & 0.15 & 0.50 & 0.20 \\ 0.15 & 0.15 & 0.10 & 0.60 \end{bmatrix}, \begin{bmatrix} 0.10 & 0.35 & 0.25 & 0.30 \\ 0.20 & 0.25 & 0.25 & 0.30 \\ 0.20 & 0.30 & 0.10 & 0.30 \\ 0.20 & 0.30 & 0.40 & 0.10 \end{bmatrix}$	0.5, 0.5

(iii) the PB rule, $\nu_{k,n}^{\text{PB}} = \frac{c_k \mu_{k,n}}{\mu_{k,N_k}}$,

(iv) the SB rue, $\nu_{k,n}^{\text{SB}} = c_k \sum_{m=1}^{n} q_{k,m}^{\text{SS}}$,

(v) the PI^{SS} rule [14] $\nu_{k,n}^{\text{SS}} = \frac{c_k \mu_{k,n}}{\sum_{m>n} q_{k,m}^{\text{SS}} (\mu_{k,m} - \mu_{k,n})}$,

where $q_{k,m}^{\text{SS}}$ is the stationary probability of state m of a user of class k.

Specifically, we only consider the case with at most one user served at each time slot. If there is more than one user having the highest index value, we uniformly choose one of them. In addition, we only consider two classes of users for a clearer performance comparison. Moreover, before evaluating the performance of different schedulers, we first test their similarity for a given scenario by computing the corresponding index and then choose one scheduler as a representative among multiple identical schedulers. In this way, we can decrease the time for numerical simulation and meanwhile obtain compact figures for performance comparison.

Let $T = 1.67$ ms for each slot for practical application [28]. The arrival probability for a new user of class k is characterized by $\xi_k = \rho_k \mu_{k,N_k}$. For comparison, we adopt the transmission rate $s_{k,\ n}$ in [28], and job size $\mathbb{E}_0\{b_k\} = 0.5\,\text{Mb}$ or HTML,

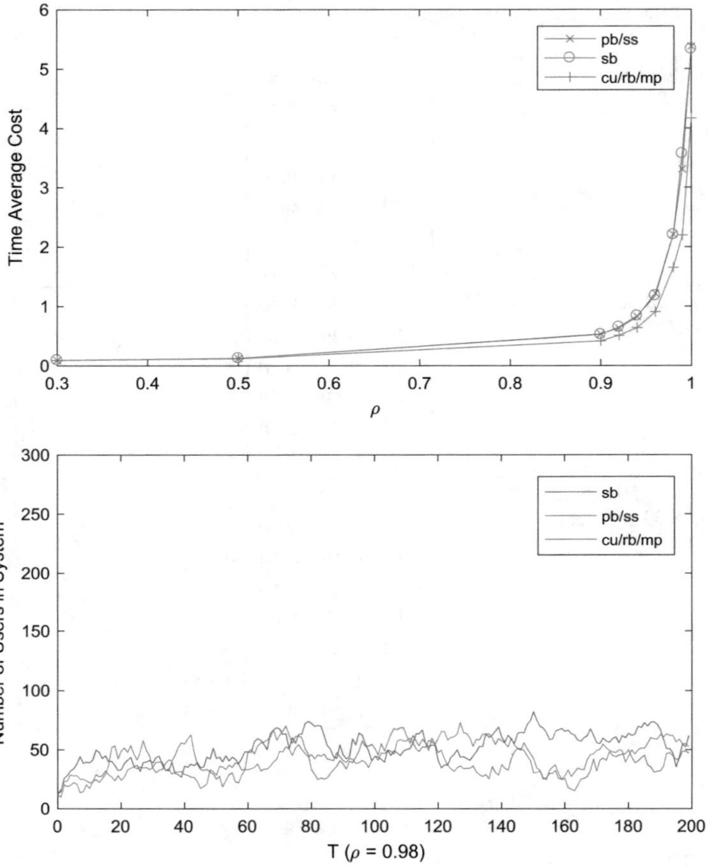

Fig. 5.1 Scenario 1 [Upper]: Time average waiting cost as a function of ρ; [Lower]: Number of users in the system as a function of time ($\rho = 0.98$)

$\mathbb{E}_0\{\mathrm{b}_k\} = 5$ Mb for PDF, and $\mathbb{E}_0\{\mathrm{b}_k\} = 50$ Mb for MP3. In this case, the departure probability is determined by $\mu_{k,n} = \tau s_{k,n}/\mathbb{E}_0\{b_k\}$. We assume that $\rho_1 = \rho_2$ and the system load $\rho = \rho_1 + \rho_2$ varies from 0.3 to 1 for a better presentation. The initial channel condition of a new user at the moment of entering the system is assumed to be determined by the stationary probability vector, i.e., with probability $q^{\mathrm{SS}}_{k,m}$ in state m for a new user of class k. The parameter setting for the following scenarios is stated in Table 5.2.

5.6.1 Scenario 1

In this case, the setting is given in Table 5.2. In particular, the users are divided into two different classes. Each user requires a job of expected size of 0:5 Mb, and has the same waiting cost $c_1 = c_2 = 1$. The channel state transition matrix is identical. But the second class of users has a better transmission rate than the first class. Our goal is to minimize the number of users waiting for service in the system.

Under this setting, three policies (*cp*, RB, and MPI) can be shown to bring about the same scheduling rule, i.e., the scheduling (class, state) order (2, 3) > (1, 3) > (1, 2) > (2, 2) > (2, 1) > (1, 1). Also, PB and $\mathrm{PI}^{\mathrm{SS}}$ yield the same result, i.e., (1, 3) = (2, 3) > (1, 2) > (2, 2) > (2, 1) > (1, 1). The SB policy will generate the order (2, 3) = (1, 3) > (1, 2) = (2, 2) > (2, 1) = (1, 1). Thus, Fig. 5.1

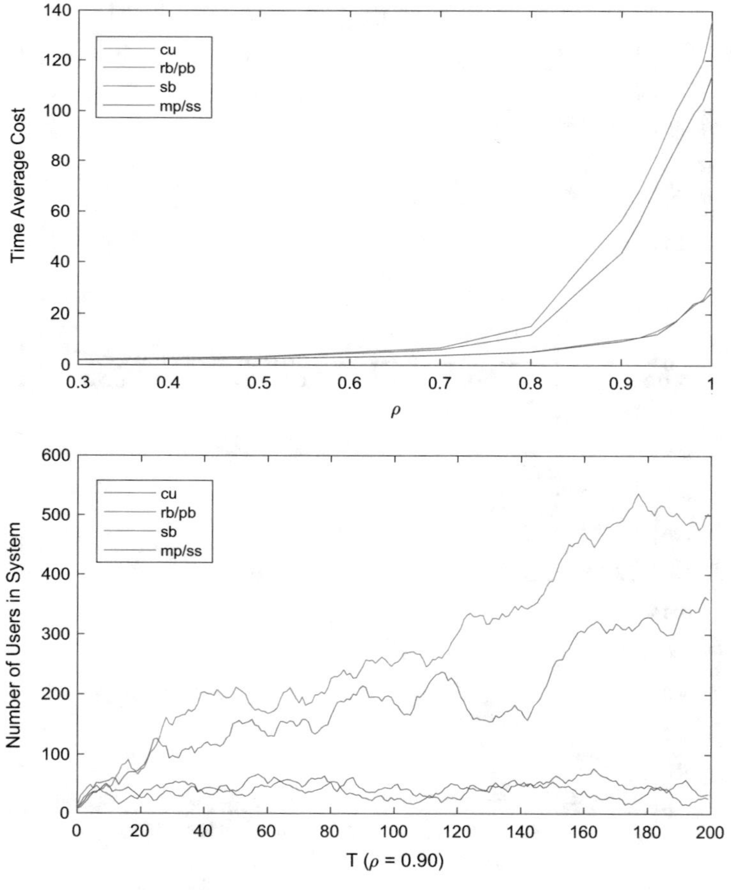

Fig. 5.2 Scenario 2 [Upper]: Time-average waiting cost as a function of ρ; [Lower]: Number of users in the system as a function of time ($\rho = 0.90$)

shows that the time-average waiting cost varies with system load ρ for three policies, and the number of users in the system varies with time slots. Obviously, we observe that the behavior of all policies is quite similar. In addition, Fig. 5.1 clearly shows that $c\mu$, RB, and MPI perform better than PB, SB, and PI^{SS}. This is because cp, RB, and MPI keep scheduling balance between class 1 and class 2. All those policies perform well with $\rho < 0.9$ but clearly have problems with stability since those policies become unstable close to 1 at which point the time-average waiting cost begins rising very steeply.

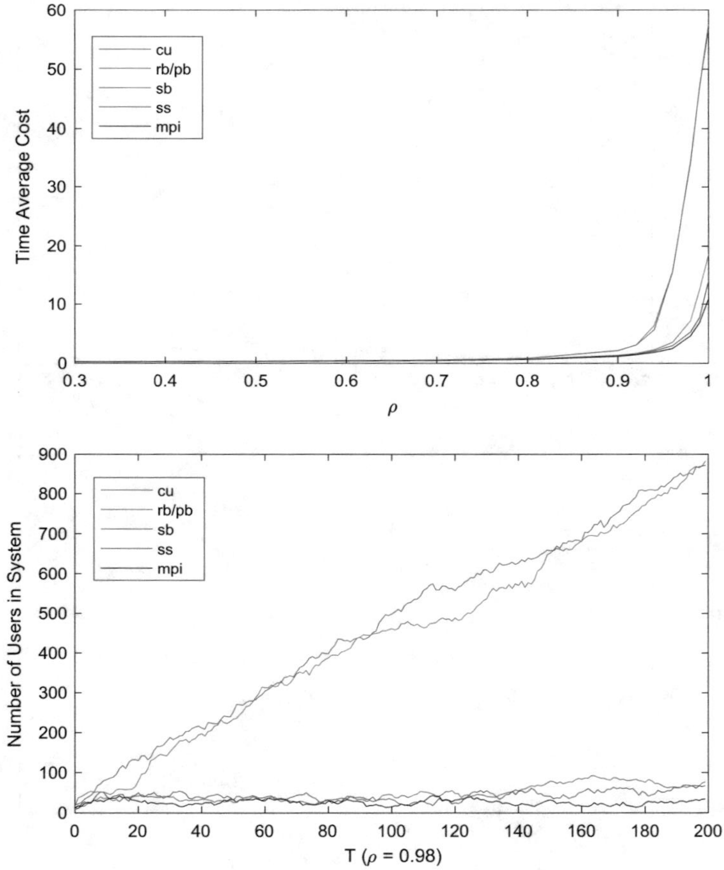

Fig. 5.3 Scenario 3 [Upper]: Time-average waiting cost as a function of ρ; [Lower]: Number of users in the system as a function of time ($\rho = 0.94$)

5.6.2 Scenario 2

In this case, we consider two classes of users with different job sizes: the first one requires a job of expected size 5 Mb, while the second one requires a job of 0.5 Mb. The waiting costs for the two classes are $c_1 = 10$ and $c_2 = 1$, respectively. The two classes have the same transmission rate and different channel state transition matrices.

Thus, we can easily check that PI^{SS} and MPI generate the same scheduling rule (1, 3) > (2, 3) > (1, 2) > (2, 2) > (1, 1) > (2, 1), and PB and RB have the same rule (1, 3) > (1, 2) > (1, 1) > (2, 3) > (2, 2) > (2, 1). Figure 5.2 shows that MPI has comparable performance with *c*/*j* and better performance than other policies in both time-average waiting cost and average number of users in system. From the scheduling order, we observe that PB (or RB) has the worst performance because of the extreme unbalance in user class, i.e., severing class 1 with complete priority than class 2. SB has the worst performance because of partial unbalance in user class from its scheduling order (1, 3) > (1, 2) > (2, 3) > (1, 1) > (2, 2) > (2, 1).

5.6.3 Scenario 3

In this case, we assume that every class of users has four states, different waiting costs, different transmission rates, and different channel transition matrices.

In this case, we can check that PB and RB are same, i.e., (1, 4) > (1, 3) > (2, 4) > (2, 3) > (1, 2) > (2, 2) > (1, 1) > (2, 1). Figure 5.3 shows that MPI policy (2, 4) > (1, 4) > (2, 3) > (1, 3) > (1, 2) > (2, 2) > (1, 1) > (2, 1) has comparable performance with PI^{SS} policy, (2, 4) = (1, 4) > (2, 3) > (1, 3) > (1, 2) > (2, 2) > (1, 1) > (2, 1), and better performance than others in both the average cost and the number of waiting users.

5.7 Summary

In this chapter, we have investigated the opportunistic scheduling problem involving multi-class multistate time-varying Markovian channels. Generally, the problem can be formulated as a restless multiarmed bandit problem. To the best of our knowledge, previous work only established an index policy for a two-state channel process and derived some limited results on multistate time-varying channels under an assumption of indexability as a prerequisite. To fill this gap, for the class of state transition matrices characterized by our proposed sufficient condition, we prove the indexability of the Whittle index policy and obtain the closed-form Whittle index. Simulation results show that the proposed index scheduler is effective in scheduling multi-class multistate channels. One future objective is to seek more generic conditions to guarantee the indexability.

Appendix

Proof of Lemma 5.1

Let v_n^* denote the optimal value function, and

$$v_n^a := r_n^a - \nu w_n^a + \beta \sum_{m \in \mathcal{N}} p_{n,m}^a v_m^*$$

$$g_n\left(v_{-n}^*, v_{-(n+1)}^*\right) := \sum_{i=1}^{n-1} \varepsilon_{N-1+i} v_i^* - \sum_{i=1}^{n-2} \varepsilon_{N-1+i} v_{i+1}^* \sum_{i=n+1}^{N-1} \varepsilon_{N-i} v_{i+1}^* - \sum_{i=n+2}^{N-1} \varepsilon_{N-i} v_i^*$$

$$\alpha_n^0 := \begin{cases} -\varepsilon_{N-n} & \text{if } n = 1 \\ -\varepsilon_{N-2+n} - \varepsilon_{N-n}\,, & \text{if } 2 \leqslant n \leqslant N-1 \\ -\varepsilon_{2N-2} & \text{if } n = N \end{cases}$$

$$\alpha_{n+1}^1 := \begin{cases} \varepsilon_{N-n} - \varepsilon_{N-n-1} & \text{if } 1 \leqslant n \leqslant N-2, \\ \varepsilon_{N-n} & \text{, if } n = N-1 \\ 0 & \text{if } n = N \end{cases}$$

For state $n \in \mathcal{N}$, the Bellman equation is

$$v_n^* = \max\left\{v_n^0; v_n^1\right\}$$

$$= \max\left\{r_n^0 - \nu w_n^0 + \beta \sum_{m \in \mathcal{N}'} h_m v_m^* + \beta g_n\left(v_{-n}^*, v_{-(n+1)}^*\right) + \beta\left[(\lambda + \alpha_n^0) v_n^* + \alpha_{n+1}^1 v_{n+1}^*\right]\right.$$

$$r_n^1 - \nu w_n^1 + \beta \sum_{m \in \mathcal{N}'} (1 - \mu_n) h_m v_m^* + \beta \mu_n v_0^* + \beta(1 - \mu_n) g_n\left(v_{-n}^*, v_{-(n+1)}^*\right)$$

$$\left. + \beta(1 - \mu_n)\left[(\lambda + \alpha_n^0) v_n^* + \alpha_{n+1}^1 v_{n+1}^*\right]\right\}$$

$$= -c + \beta \sum_{m \in \mathcal{N}^*} h_m v_m^* + \beta g_n\left(v_{-n}^*, v_{-(n+1)}^*\right) + \beta\left[\left(\lambda + \alpha_n^0\right) v_n^* + \alpha_{n+1}^1 v_{n+1}^*\right]$$

$$+ \max\left\{0; -\nu + \mu_n\left(c + \beta v_0^* - \beta \sum_{m \in \mathcal{N}'} h_m v_m^* - \beta g_n\left(v_{-n}^*, v_{-(n+1)}^*\right)\right.\right.$$

$$\left.\left. - \beta\left[(\lambda + \alpha_n^0) v_n^* + \alpha_{n+1}^1 v_{n+1}^*\right]\right)\right\}$$

where the first term in the curly brackets correspond to action 0 and the second to action 1.

Obviously, transmitting (i.e., action 1) is optimal in state $n \in \mathcal{N} \backslash \{0\}$ if the first term is less than the second one.

For ease of presentation, let

$$Z{:=}c+\beta v_0^*-\beta g_n\left(v_{-n}^*,v_{-(n+1)}^*\right)-\beta\sum_{m\in\mathcal{N}'}^{m\neq n,n+1}h_m v_m^*$$

$$Z_n{:=}Z-\beta\left(\lambda+\alpha_n^0+h_n\right)v_n^*-\beta\left(\alpha_{n+1}^1+h_{n+1}\right)v_{n+1}^*$$

Now, we analyze the Bellman equation by two cases.

Case 1. If $\mathbf{v}>0$, we have $v_n^*\leqslant 0$ for any $n\in\mathcal{N}\backslash\{0\}$.I. If transmitting is optimal in state $n\in\mathcal{N}\backslash\{0,N\}$, we obtain $-v+\mu_n Z_n\geqslant 0$ indicating $Z_n>0$, and further, $-v+\mu_{n+1}Z_n>-v+\mu_n Z_n$ since $\mu_{n+1}>\mu_n$. Thus

$$v_n^*=-c+\mu_n Z_n-\nu+\beta\sum_{m\in\mathcal{N}'}h_m v_m^*+\beta\left[\left(\lambda+\alpha_n^0\right)v_n^*+\alpha_{n+1}^1 v_{n+1}^*+g_n\left(v_{-n}^*,v_{-(n+1)}^*\right)\right]$$

$$<-c+\mu_{n+1}Z_n-\nu+\beta\sum_{m\in\mathcal{N}'}h_m v_m^*+\beta\left[\left(\lambda+\alpha_n^0\right)v_n^*+\alpha_{n+1}^1 v_{n+1}^*+g_n\left(v_{-n}^*,v_{-(n+1)}^*\right)\right],$$

equivalently,

$$v_n^*\left(1-\beta\left(1-\mu_{n+1}\right)\left(\lambda-\varepsilon_{N-n}-\varepsilon_{N+n-1}\right)\right)$$

$$<-c+\mu_{n+1}Z-\nu+\beta\sum_{m\in\mathcal{N}'}^{m\neq n,n+1}h_m v_m^*+\beta g_n\left(v_{-n}^*,v_{-(n+1)}^*\right)$$

$$+\beta\left(1-\mu_{n+1}\right)\left(h_n+\alpha_n^0+\varepsilon_{N-n}+\varepsilon_{N+n-1}\right)v_n^*+\beta\left(1-\mu_{n+1}\right)\left(h_{n+1}+\alpha_{n+1}^1\right)v_{n+1}^*$$

$$=-c+\mu_{n+1}Z-\nu+\beta\sum_{m\in\mathcal{N}'}^{m\neq n,n+1}h_m v_m^*+\beta g_n\left(v_{-n}^*,v_{-(n+1)}^*\right)$$

$$+\beta\left(1-\mu_{n+1}\right)\left(h_n+\gamma_n\right)v_n^*+\beta\left(1-\mu_{n+1}\right)\left(h_{n+1}+\alpha_{n+1}^1\right)v_{n+1}^* \tag{5.41}$$

where, $\gamma_n{:=}\alpha_n^0+\varepsilon_{N+n-1}+\varepsilon_{N-n}$

For state $n+1$, if action "1" is adopted, then we have according to Bellman equation:

$$v_{n+1}^1=-c-\nu+\beta\sum_{m\in\mathcal{N}'}h_m v_m^*+\beta g_{n+1}\left(v_{-(n+1)}^*,v_{-(n+2)}^*\right)$$

$$+\beta\left[\left(\lambda+\alpha_{n+1}^0\right)v_{n+1}^*+\alpha_{n+2}^1 v_{n+2}^*\right]+\mu_{n+1}\left(c+\beta v_0^*-\beta\sum_{m\in\mathcal{N}'}h_m v_m^*\right.$$

$$\left.-\beta g_{n+1}\left(v_{-(n+1)}^*,v_{-(n+2)}^*\right)-\beta\left[\left(\lambda+\alpha_{n+1}^0\right)v_{n+1}^*+\alpha_{n+2}^1 v_{n+2}^*\right]\right)$$

$$
\begin{aligned}
&\overset{(a)}{=} -c + \mu_{n+1} Z - \nu + \beta \sum_{m \in \mathcal{N}'}^{m \neq n, n+1} h_m v_m^* + \beta g_n(v_{-n}^*, v_{-(n+1)}^*) \\
&+\beta\big(1-\mu_{n+1}\big)\big(h_n + \alpha_n^0 + \varepsilon_{N+n-1} + \varepsilon_{N-n}\big) v_n^* \\
&+\beta\big(1-\mu_{n+1}\big)\big(h_{n+1} + \lambda + \alpha_{n+1}^1 - \varepsilon_{N-n} - \varepsilon_{N+n-1}\big) v_{n+1}^* \\
&\quad \Leftrightarrow \\
&\quad v_{n+1}^1 - \beta\big(1-\mu_{n+1}\big)(\lambda - \varepsilon_{N-n} - \varepsilon_{N+n-1}) v_{n+1}^* \\
&\qquad = -c + \mu_{n+1} Z - \nu + \beta \sum_{m \in \mathcal{N}'}^{m \neq \eta, n+1} h_m v_m^* \\
&\quad +\beta g_n\Big(v_{-n}^*, v_{-(n+1)}^*\Big) + \beta\big(1-\mu_{n+1}\big)(h_n + \gamma_n) v_n^* \\
&\quad +\beta\big(1-\mu_{n+1}\big)\big(h_{n+1} + \alpha_{n+1}^1\big) v_{n+1}^*
\end{aligned}
\tag{5.42}
$$

where (a) is due to $g_n\Big(v_{-n}^*, v_{-(n+1)}^*\Big) = g_{n+1}\Big(v_{-(n+1)}^*, v_{-(n+2)}^*\Big) + (\varepsilon_{N-2+n} - \varepsilon_{N-1+n}) v_n^* + (\varepsilon_{N-n-1} - \varepsilon_{N-n-2}) v_{n+2}^*, \alpha_{n+1}^0 = \alpha_{n+1}^1 - \varepsilon_{N-n} - \varepsilon_{N+n-1}$ and $\alpha_{n+2}^1 = \varepsilon_{N-n-1} - \mathcal{E}_N - n - 2$.

Thus, combining (5.41) and (5.42), we have

$$
\begin{aligned}
&v_n^*\big(1 - \beta\big(1-\mu_{n+1}\big)(\lambda - \varepsilon_{N-n} - \varepsilon_{N+n-1})\big) \\
&< v_{n+1}^1 - \beta\big(1-\mu_{n+1}\big)(\lambda - \varepsilon_{N-n} - \varepsilon_{N+n-1}) v_{n+1}^* \\
&\leqslant v_{n+1}^* - \beta\big(1-\mu_{n+1}\big)(\lambda - \varepsilon_{N-n} - \varepsilon_{N+n-1}) v_{n+1}^* \\
&= v_{n+1}^*\big(1 - \beta\big(1-\mu_{n+1}\big)(\lambda - \varepsilon_{N-n} - \varepsilon_{N+n-1})\big)
\end{aligned}
$$

which indicates $v_n^* < v_{n+1}^*$.

Meanwhile,

$$
\begin{aligned}
&v_n^* \geqslant v_n^0 \\
&= -c + \beta \sum_{m \in \mathcal{N}'} h_m v_m^* + \beta g_n\Big(v_{-n}^*, v_{-(n+1)}^*\Big) + \beta\Big[\big(\lambda + \alpha_n^0\big) v_n^* + \alpha_{n+1}^1 v_{n+1}^*\Big] \\
&= -c + \beta \sum_{m \in \mathcal{N}'}^{m \neq n, n+1} h_m v_m^* + \beta g_n\Big(v_{-n}^*, v_{-(n+1)}^*\Big) \\
&\quad + \beta\Big[\big(h_n + \lambda + \alpha_n^0\big) v_n^* + \big(h_{n+1} + \alpha_{n+1}^1\big) v_{n+1}^*\Big] \\
&\Leftrightarrow \\
&v_n^*(1 - \beta(\lambda - \varepsilon_{N-n} - \varepsilon_{N+n-1}))
\end{aligned}
$$

$$\geqslant -c+\beta \sum_{m\in\mathcal{N}'}^{m\neq n,n+1} h_m v_m^* + \beta g_n\left(v_{-n}^*, v_{-(n+1)}^*\right)$$
$$+\beta\left[(h_n+\gamma_n)v_n^* + \left(h_{n+1}+\alpha_{n+1}^1\right)v_{n+1}^*\right] \tag{5.43}$$

On the other hand, we have according to Bellman equation

$$v_{n+1}^0 = -c+\beta\sum_{m\in\mathcal{N}'} h_m v_m^* + \beta g_{n+1}\left(v_{-(n+1)}^*, v_{-(n+2)}^*\right)$$
$$+\beta\left[\left(\lambda+\alpha_{n+1}^0\right)v_{n+1}^* + \alpha_{n+2}^1 v_{n+2}^*\right]$$
$$\overset{(b)}{=} -c+\beta\sum_{m\in\mathcal{N}'} h_m v_m^* + \beta g_n\left(v_{-n}^*, v_{-(n+1)}^*\right)$$
$$+\beta\left[\gamma_n v_n^* + \left(\lambda+\alpha_{n+1}^1 - \varepsilon_{N-n} - \varepsilon_{N+n-1}\right)v_{n+1}^*\right]$$
$$= -c+\beta\sum_{m\in\mathcal{N}'}^{m\neq\eta,n+1} h_m v_m^* + \beta g_n\left(v_{-n}^*, v_{-(n+1)}^*\right)$$
$$+\beta\left[(h_n+\gamma_n)v_n^* + \left(h_{n+1}+\alpha_{n+1}^1\right)v_{n+1}^*\right] + \beta(\lambda-\varepsilon_{N-n}-\varepsilon_{N+n-1})v_{n+1}^*$$
$$\Leftrightarrow$$
$$v_{n+1}^0(1-\beta(\lambda-\varepsilon_{N-n}-\varepsilon_{N+n-1})$$
$$\leqslant -c+\beta\sum^{m\neq n,n+1} h_m v_m^* + \beta g_n\left(v_{-n}^*, v_{-(n+1)}^*\right)$$
$$+\beta\left[(h_n+\gamma_n)v_n^* + \left(h_{n+1}+\alpha_{n+1}^1\right)v_{n+1}^*\right] \tag{5.44}$$

where (a) is due to $\lambda \leqslant \varepsilon_{N-n} + \varepsilon_{N+n-1}$ and $v_{n+1}^0 \leqslant v_{n+1}^*$, and (b) is due to $g_n\left(v_{-n}^*, v_{-(n+1)}^*\right) = g_{n+1}\left(v_{-(n+1)}^*, v_{-(n+2)}^*\right) + \gamma_n v_n^* + (\varepsilon_{N-n-1} - \varepsilon_{N-n-2})v_{n+2}^*$.

Thus, combining (5.43) and (5.44), we have $v_{n+1}^0 \leqslant v_n^*$. Since $v_n^* < v_{n+1}^* g_{n+1}\left(v_{-(n+1)}^*, v_{-(n+2)}^*\right) + \gamma_n v_n^* + (\varepsilon_{N-n-1} - \varepsilon_{N-n-2})v_{n+2}^*$, we conclude $v_{n+1}^0 \leqslant v_n^* < v_{n+1}^* = v_{n+1}^1$, that is, transmitting is optimal in state n + 1.

Case 2. if $v < 0$, then we proceed as follows. First, using the Bellman equation it is easy to obtain that $v_0^* = -\frac{\nu}{1-\beta}$ because action 1 is optimal in state 0 and thus $-v$ is obtained in every period forever. Notice that the one-period net reward, $r_n^a - \nu w_n^a$, is for any state $n \in \mathcal{N}$ and any action a G A upperbounded by –i.e., $\mid r_n^a - \nu r_n^a \mid \leqslant -\nu$. Hence, $v_m^* \leqslant -\frac{\nu}{1-\beta} = v_0^*$ for any m G N', and therefore (using $c > 0$ and $\lambda +$

$\sum_{m\in\mathcal{N}'} h_m = 1)$ $Z_n > 0$, and finally, for any state $n \in \mathcal{N} \setminus \{0\}$, $-\mathrm{v} + /\mu_n Z_n > 0$. That is, transmitting is optimal in any state $n \in \mathcal{N}$.

Combining the two cases, we complete the proof.

Proof of Lemma 5.2

We first show $\mathbb{W}_n^{\langle 1,\delta_{N-n}\rangle} - \mathbb{W}_n^{\langle 0,\delta_{N-n}\rangle} > 0$ by the following four steps.

Step 1. According to the definition of β-average work, we have $\mathbb{W}_n^{1,\delta_{N-n}} \geqslant 1$, $\mathbb{W}_{n+1}^{\langle 1,\delta_{N-n}\rangle} \geqslant 1, \cdots, \mathbb{W}_N^{\langle 1,\delta_{N-n}\rangle} \geqslant 1$ To show $\mathbb{W}_n^{\langle 1,\delta_{N-n}\rangle} \geqslant \mathbb{W}_{n+1}^{\langle 1,\delta_{N-n}\rangle} \geqslant \cdots \geqslant \mathbb{W}_N^{\langle 1,\delta_{N-n}\rangle}$, we only need to show $\mathbb{W}_i^{\langle 1,\delta_{N-n}\rangle} \geqslant \mathbb{W}_{i+1}^{\langle 1,\delta_{N-n}\rangle}$ for any $i(n \leqslant i \leqslant N-1)$.

For (5.24), we perform the following operations sequentially:

(i) Dividing the i-th equation by $1 - \mu_i$,
(ii) Dividing the $(i + 1)$-th equation by,
(iii) Subtracting the i-th equation from the $(i + 1)$-th one,

then we obtain the i-th equation

$$\left[-\frac{1}{\bar{\mu}_i} + \beta\lambda_{i-1}\right]\mathbb{W}_i^{\langle 1,\delta_{N-n}\rangle} + \left[\frac{1}{\bar{\mu}_{i+1}} - \beta\lambda_{i-1}\right]\mathbb{W}_{i+1}^{\langle 1,\delta_{N-n}\rangle} = \frac{1}{\bar{\mu}_{i+1}} - \frac{1}{\bar{\mu}_i}$$

equivalently,

$$\left[\frac{1}{\bar{\mu}_i} - \beta\lambda_{i-1}\right]\left(\mathbb{W}_{i+1}^{\langle 1,\delta_{N-n}\rangle} - \mathbb{W}_i^{\langle 1,\delta_{N-n}\rangle}\right) = \left[\frac{1}{\bar{\mu}_{i+1}} - \frac{1}{\bar{\mu}_i}\right]\left(1 - \mathbb{W}_{i+1}^{\langle 1,\delta_{N-n}\rangle}\right) \leqslant 0$$

which implies $\mathbb{W}_{i+1}^{\langle 1,\delta_{N-n}\rangle} \leqslant \mathbb{W}_j^{\langle 1,\delta_{N-n}\rangle}$.

Step 2 To show $\mathbb{W}_1^{\langle 0,\delta_{N-n}\rangle} = \cdots = \mathbb{W}_{n-1}^{\langle 0,\delta_{N-n}\rangle}$ we only need to show $\mathbb{W}_i^{\langle 0,\delta_{N-n}\rangle} = \mathbb{W}_{i+1}^{\langle 0,\delta_{N-n}\rangle} \mathbb{W}^{0,\delta_{N-n}}$ for $i(1 \leqslant i \leqslant n-2)$.

For (5.24), we subtract the i-th equation from the $(i + 1)$-th one, and come to

$$-[1 - \beta\lambda_i]\mathbb{W}_i^{\langle 0,\delta_{N-n}\rangle} + [1 - \beta\lambda_i]\mathbb{W}_{i+1}^{\langle 0,\delta_{N-n}\rangle} = 0 \tag{5.46}$$

which indicates$\mathbb{W}_i^{\langle 0,\delta_{N-n}\rangle} = \mathbb{W}_{i+1}^{\langle 0,\delta_{N-n}\rangle}$.

To show $\mathbb{W}_1^{\langle 0,\delta_{N-n}\rangle} = \cdots = \mathbb{W}_{n-1}^{\langle 0,\delta_{N-n}\rangle} \leqslant \mathbb{W}_n^{\langle 1,\delta_{N-n}\rangle}$, we have by the $(n - 1)$-th equation

$$[1 - \beta\sum_{i=1}^{n-1} q_{n-1,i}]\mathbb{W}_{n-1}^{\langle 0,\delta_{N-n}\rangle} - \beta\sum_{i=n}^{N} q_{n-1,i}\mathbb{W}_i^{\langle 1,\delta_{N-n}\rangle} = 0 \tag{5.47}$$

equivalently,

$$\begin{aligned}
\mathbb{W}_{n-1}^{\langle 0,\delta_{N-n}\rangle} &= \frac{\beta \sum_{i=n}^{N} q_{n-1,i} \mathbb{W}_{i}^{\langle 1,\delta_{N-n}\rangle}}{1-\beta \sum_{i=1}^{n-1} q_{n-1,i}} \\
&\leqslant \frac{\beta \sum_{i=n}^{N} q_{n-1,i} \mathbb{W}_{n}^{\langle 1,\delta_{N-n}\rangle}}{1-\beta \sum_{i=1}^{n-1} q_{n-1,i}} \\
&\leqslant \frac{\sum_{i=n}^{N} q_{n-1,i} \mathbb{W}_{n}^{\langle 1,\delta_{N-n}\rangle}}{1- \sum_{i=1}^{n-1} q_{n-1,i}} \\
&= \mathbb{W}_{n}^{\langle 1,\delta_{N-n}\rangle} \qquad (5.48)
\end{aligned}$$

where (a) is due to $\mathbb{W}_{n}^{\langle 1,\delta_{N-n}\rangle} \geqslant \mathbb{W}_{n+1}^{\langle 1,\delta_{N-n}\rangle} \geqslant \cdots \geqslant \mathbb{W}_{N}^{\langle 1,\delta_{N-n}\rangle}$, and (b) is because $\frac{\beta \sum_{i=n}^{N} q_{n-1,i}}{1-\beta \sum_{i=1}^{n-1} q_{n-1,i}}$ is increasing in β $(0 < \beta < 1)$.

Step 3. Considering the n-th equation of (5.24), we have

$$\begin{aligned}
-\beta(1-\mu_n)\sum_{i=1}^{n-1} q_{n,i}\mathbb{W}_{n-1}^{\langle 0,\delta_{N-n}\rangle} + (1-\beta(1-\mu_n)q_{n,n})\mathbb{W}_{n}^{\langle 1,\delta_{N-n}\rangle} \\
-\beta(1-\mu_n)\sum_{i=n+1}^{N} q_{n,i}\mathbb{W}_{i}^{\langle 1,\delta_{N-n}\rangle} = 1 \qquad (5.49)
\end{aligned}$$

equivalently,

$$\begin{aligned}
&(1-\beta(1-\mu_n)q_{n,n})\mathbb{W}_{n}^{\langle 1,\delta_{N-n}\rangle} \\
&= 1+\beta(1-\mu_n)\sum_{i=1}^{n-1} q_{n,i}\mathbb{W}_{n-1}^{\langle 0,\delta_{N-n}\rangle} + \beta(1-\mu_n)\sum_{i=n+1}^{N} q_{n,i}\mathbb{W}_{i}^{\langle 1,\delta_{N-n}\rangle} \\
&\overset{(a)}{\leqslant} 1+\beta(1-\mu_n)\sum_{i=1}^{n-1} q_{n,i}\mathbb{W}_{n}^{\langle 1,\delta_{N-n}\rangle} + \beta(1-\mu_n)\sum_{i=n+1}^{N} q_{n,i}\mathbb{W}_{n}^{\langle 1,\delta_{N-n}\rangle} \qquad (5.50)
\end{aligned}$$

further,

$$\mathbb{W}_n^{\langle 1,\delta_{N-n}\rangle} \leqslant \frac{1}{1-\beta(1-\mu_n)} < \frac{1}{\mu_n} \tag{5.51}$$

due to, and (a) is due to $\mathbb{W}_n^{\langle 1,\delta_{N-n}\rangle} \geqslant \mathbb{W}_i^{\langle 0,\delta_{N-n}\rangle}$ for any $i(1 \leqslant i \leqslant n-1)$ and $\mathbb{W}_n^{\langle 1,\delta_{N-n}\rangle} \geqslant \mathbb{W}_i^{\langle 1,\delta_{N-n}\rangle}$ for any $i(n+1 \leqslant i \leqslant N)$.

Step 4. For the n-th equation of (5.24) is stated as follows:

$$\left(\boldsymbol{e}_n^\top - \beta(1-\mu_n)\boldsymbol{e}_n^\top \boldsymbol{Q}\right)\boldsymbol{w}_1 = 1 \tag{5.52}$$

we first subtract $\mu_n\mathbb{W}_n^{\langle 1,\delta_{N-n}\rangle}$ from both LHS and RHS of (5.52), and then divide (5.52) by $1-\mu_n$. As a consequence, (5.52) can be written as follows:

$$(\boldsymbol{e}_n^\top - \beta\boldsymbol{e}_n^\top \boldsymbol{Q})\boldsymbol{w}_1 = \frac{1-\mu_n\mathbb{W}_n^{\langle 1,\delta_{N-n}\rangle}}{1-\mu_n} \tag{5.53}$$

Combined with the other N - 1 equations of (5.24), then (5.24) can be transformed to the following:

$$\begin{aligned} &(\boldsymbol{I}_N - \beta\boldsymbol{M}_1)\cdot\boldsymbol{w}_1 = \boldsymbol{1}_N - \boldsymbol{1}_N^{n-1} \\ \Leftrightarrow &(\boldsymbol{I}_N - \beta\boldsymbol{M}_0)\cdot\boldsymbol{w}_1 = \boldsymbol{1}_N - \boldsymbol{1}_N^{n} + \frac{1-\mu_n\mathbb{W}_n^{\langle 1,\delta_{N-n}\rangle}}{1-\mu_n}\boldsymbol{e}_n \end{aligned} \tag{5.54}$$

Thus,

$$\mathbb{W}_n^{\langle 1,\delta_{N-n}\rangle} = \boldsymbol{e}_n^\top(\boldsymbol{I}_N - \beta\boldsymbol{M}_0)^{-1}(1_N - 1_N^n + \frac{1-\mu_n\mathbb{W}_n^{\langle 1,\delta_{N-n}\rangle}}{1-\mu_n}\boldsymbol{e}_n) \tag{5.55}$$

combined with

$$\mathbb{W}_n^{\langle 0,\delta_{N-n}\rangle} = \boldsymbol{e}_n^\top(\boldsymbol{I}_N - \beta\boldsymbol{M}_0)^{-1}(\boldsymbol{1}_N - \boldsymbol{1}_N^n)$$

then we have

$$\mathbb{W}_n^{\langle 1,\delta_{N-n}\rangle} - \mathbb{W}_n^{\langle 0,\delta_{N-n}\rangle} = \frac{1-\mu_n\mathbb{W}_n^{\langle 1,\delta_{N-n}\rangle}}{1-\mu_n}[\boldsymbol{e}_n^\top(\boldsymbol{I}_N - \beta\boldsymbol{M}_0)^{-1}\boldsymbol{e}_n] \overset{(a)}{>} 0 \tag{5.56}$$

where (a) is due to $\mu_n\mathbb{W}_n^{\langle 1,\delta_{N-n}\rangle} < 1$ and $\boldsymbol{e}_n^\top(\boldsymbol{I}_N - \beta\boldsymbol{M}_0)^{-1}\boldsymbol{e}_n > 0$. Note that $\boldsymbol{e}_n^\top(\boldsymbol{I}_N - BM_0)^{-1}e_n > 0$ because $I_N - \beta M_0$ is a diagonally dominant matrix, and every element in the diagonal line is larger than 0 when $0 < \beta < 1$.

To the end, we prove $\mathbb{W}_n^{\langle 1,\delta_{N-n}\rangle} - \mathbb{W}_n^{\langle 0,\delta_{N-n}\rangle} > 0$. Following the similar deduction, we can easily prove $\mathbb{W}_{n+1}^{\langle 1,\delta_{N-n}\rangle} - \mathbb{W}_{n+1}^{\langle 0,\delta_{N-n}\rangle} > 0$. Therefore, we complete the proof.

Proof of Theorem 5.2

According to the definition of the Whittle index, we prove the indexability by checking $v_1 < v_2 < \cdots < v_N$. When $\beta = 1$, we have $\nu_N = \frac{c\mu_N}{1-\beta} \to \infty$.

First, we have

$$f(n) = \beta q_{n,n+1}\left(1 - \frac{\frac{1}{\mu_n} - \beta\lambda_n}{\frac{1}{\mu_{n+1}} - \beta\lambda_n}\right) + \mu_n d_{n-1,n+1} K_{n+1}$$

$$+\beta \sum_{i=n+2}^{N} q_{n,i}\left(1 - \prod_{j=n+1}^{i} \frac{\frac{1}{\mu_{j-1}} - \beta\lambda_{j-1}}{\frac{1}{\mu_j} - \beta\lambda_{j-1}}\right) + \mu_n \sum_{i=n+2}^{N} d_{n-1,i} K_i$$

And

$$f(n+1) = \beta \sum_{i=n+2}^{N} q_{n+1,i}\left(1 - \prod_{j=n+2}^{i} \frac{\frac{1}{\mu_{j-1}} - \beta\lambda_{j-1}}{\frac{1}{\mu_j} - \beta\lambda_{j-1}}\right) + \mu_{n+1} \sum_{i=n+2}^{N} d_{n,i} K_i$$

$$= \beta \sum_{i=n+2}^{N} q_{n,i}\left(1 - \prod_{j=n+2}^{i} \frac{\frac{1}{\mu_{j-1}} - \beta\lambda_{j-1}}{\frac{1}{\mu_j} - \beta\lambda_{j-1}}\right) + \mu_{n+1} \sum_{i=n+2}^{N} d_{n-1,i} K_i$$

Further,

$$f(n) - f(n+1)$$

$$= \beta q_{n,n+1} K_{n+1} + \left(\mu_n - \mu_{n+1}\right) \sum_{i=n+2}^{N} d_{n-1,i} K_i$$

$$+\mu_n d_{n-1,n+1} K_{n+1} + \beta \sum_{i=n+2}^{N} q_{n,i} \prod_{j=n+2}^{i} \frac{\frac{1}{\mu_{j-1}} - \beta\lambda_{j-1}}{\frac{1}{\mu_j} - \beta\lambda_{j-1}} K_{n+1}$$

$$= \beta q_{n,n+1} K_{n+1} + \left(\mu_n - \mu_{n+1}\right) \sum_{i=n+2}^{N} d_{n-1,i} K_i$$

$$-\mu_n \beta\left(q_{n-1,n+1} + \sum_{i=n+2}^{N} q_{n-1,i} \prod_{j=n+2}^{i} \frac{\frac{1}{\mu_{j-1}} - \beta\lambda_{j-1}}{\frac{1}{\mu_j} - \beta\lambda_{j-1}}\right) K_{n+1}$$

$$+\beta \sum_{i=n+2}^{N} q_{n,i} \prod_{j=n+2}^{i} \frac{\frac{1}{\mu_{j-1}} - \beta\lambda_{j-1}}{\frac{1}{\mu_j} - \beta\lambda_{j-1}} K_{n+1}$$

$$= \beta q_{n,n+1} K_{n+1} + (\mu_n - \mu_{n+1}) \sum_{i=n+2}^{N} d_{n-1,i} K_i$$

$$-\mu_n \beta \left(q_{n,n+1} + \sum_{i=n+2}^{N} q_{n,i} \prod_{j=n+2}^{i} \frac{\frac{1}{\mu_{j-1}} - \beta\lambda_{j-1}}{\frac{1}{\overline{\mu}_j} - \beta\lambda_{j-1}} \right) K_{n+1}$$

$$\mathrm{b}+\beta \sum_{i=n+2}^{N} q_{n,i} \prod_{j=n+2}^{i} \frac{\frac{1}{\mu_{j-1}} - \beta\lambda_{j-1}}{\frac{1}{\mu_j} - \beta\lambda_{j-1}} K_{n+1}.$$

$$= \overline{\mu}_n \beta q_{n,n+1} K_{n+1} + (\mu_n - \mu_{n+1}) \sum_{i=n+2}^{N} d_{n-1,i} K_i$$

$$+\overline{\mu}_n \beta \sum_{i=n+2}^{N} q_{n,i} \prod_{j=n+2}^{i} \frac{\frac{1}{\mu_{j-1}} - \beta\lambda_{j-1}}{\frac{1}{\mu_j} - \beta\lambda_{j-1}} K_{n+1}$$

$$\overset{(a)}{\geqslant} 0,$$

where (a) is due to $d_{n-1,i} \leqslant 0$, $K_i \geqslant 0$, and $\mu_n < \mu_{n+1}$. Hence,

$$v_n = \frac{c\mu_n}{1 - \beta + f(n)} \leqslant \frac{c\mu_n}{1 - \beta + f(n+1)} < \frac{c\mu_{n+1}}{1 - \beta + f(n+1)} = v_{n+1}.$$

which complete the proof.

References

1. K. Wang, L.Chen, J. Yu, M.Z. Win. Opportunistic Scheduling Revisited Using Restless Bandits: Indexability and Index Policy. in *Proc. of GlobeCom*, Singapore, 1–6 Dec 2017
2. R. Knopp, P. Humblet, Information capacity and power control in single-cell multiuser communications. in *Proc. IEEE Int. Conf. Commun.*, Seattle, WA, Mar 1995, pp. 331–33
3. P. Bender, P. Black, M. Grob, R. Padovani, N. Sindhushayana, A. Viterbi, CDMA/HD a bandwidth-efficient high-speed wireless data service for nomadic users. IEEE Comm Mag. **38** (7), 70–77 (2000)
4. H.J. Kushner, P.A. Whiting, Convergence of proportional fair sharing algorithms in general conditions. IEEE Trans. Wireless Commun. **3**(4), 1250–1259 (2004)
5. S. Borst, User-level performance of channel-aware scheduling algorithms in wireless d networks. IEEE/ACM Trans. Netw. **13**(3), 636–647 (2005)
6. S. Aalto, P. Lassila, Flow-level stability and performance of channel-aware priority-bas schedulers. in *6th EURO-NGI Conference on Next Generation Internet*, Paris, France, 20, pp. 1–8
7. T. Bonald, A score-based opportunistic scheduler for fading radio channels. in *Proc. of European Wireless*, 2004, pp. 283–292
8. U. Ayesta, M. Erausquin, M. Jonckheere, I.M. Verloop, Scheduling in a random environment: Stability and asymptotic optimality. IEEE/ACM Trans. Netw. **21**(1), 258–271 (2013)

9. J. Kim, B. Kim, J. Kim, Y.H. Bae, Stability of flow-level scheduling with Markovian time-varying channels. Perform. Eval. **70**(2), 148–159 (2013)
10. S. Aalto, P. Lassila, P. Osti, Whittle index approach to size-aware scheduling with time varying channels. in *Proc.* ACM Sigmetrics, Portland, OR, June 2015
11. S. Aalto, P. Lassila, P. Osti, Whittle index approach to size-aware scheduling for time-varying channels with m tiple states. Queueing Syst. **83**, 195–225 (2016)
12. F. Cecchi P. Jacko, Scheduling of Users with Markovian Time-Varying Transmission Rate, in *Proc. ACM Sigmetrics*, Pittsburgh, PA, 2013
13. P. Jacko, Value of information in optimal flow-level scheduling of users with Markovian time varying channels. Perform. Eval. **68**(11), 1022–1036 (2011)
14. F. Cecchi, P. Jacko, Nearly-optimal scheduling of users with Markovian time-varying transmission rates. Perform. Eval. **99–100**, 16–36 (2016)
15. K. Wang, Q. Liu, Q. Fan, Q. Ai, Optimally probing channel in opportunistic communication access. IEEE Commun. Lett. **22**(7), 1426–1429 (2018)
16. K. Wang, L. Chen, J. Yu, Q. Fan, Y. Zhang, W. Chen, P. Zhou, Y. Zhong, On optimal of second-highest policy for opportunistic multichannel access. IEEE Trans. Veh. Techn **22**(7), 1426–1429 (2018)
17. P. Whittle, Restless bandits: activity allocation in a changing world. J. Appl. Probab. **25A**, 287–298 (1988)
18. U. Ayesta, M. Erausquin, P. Jacko, A modeling framework for optimizing the flow-le scheduling with time-varying channels. Perform. Eval. **67**, 1014–1029 (2010)
19. P. Jacko, E. Morozov, L. Potakhina, I. M. Verloop, Maximal flow-level stability of be rate schedulers in heterogeneous wireless systems, Trans. Emerg. Telecommun. Technol. (2015);28:e2930
20. I. Taboada, F. Liberal, P. Jacko, An opportunistic and non-anticipating size-aware scheduling proposal for mean holding cost minimization in time-varying channels. Perform. Eval. **79**, 90–103 (2014)
21. I. Taboada, P. Jacko, U. Ayesta, F. Liberal, Opportunistic scheduling of flows with general size distribution in wireless time-varying channels, in *Proc. of ITC-26*, 2014
22. S. Aalto, P. Lassila, P. Osti, On the optimal trade-off between SRPT and opportunistic scheduling, in *Proc. ACM Sigmetrics*, San Jose, CA, June 2011
23. T. Bonald, S. Borst, N. Hegde, M. Jonckheere, A. Proutiere, Flow-level performance and capacity of wireless networks with user mobility. Queueing Syst. **63**, 131–164 (2009)
24. S. Borst, M. Jonckheere, Flow-level stability of channel-aware scheduling algorithms, in *Proc. of WiOpt*, 2006
25. P. van de Ven, S. Borst, S. Shneer, Instability of maxweight scheduling algorithms, in *Proc. IEEE Conf. on Computer Commun.*, 2009, pp. 1701–1709
26. C. Buyukkoc, P. Varaiya, J. Walrand, The $c\beta$ rule revisited. Adv. Appl. Probab. **17**(1), 237–238 (1985)
27. J. Nino-Mora, Characterization and computation of restless bandit marginal productivity indices, in *Proc. of ValueTools*, 2007
28. S. Sesia, I. Toufik, M. Baker, *LTE-the UMTS Long Term Evolution: From Theory to Practice* (John Wiley & Sons Ltd., New York, NY, 2011)

Chapter 6
Conclusion and Perspective

6.1 Summary

This book addresses a special kind of restless multiarmed bandit problem arising in opportunistic scheduling with imperfect sensing or observation conditions where each channel evolves as a discrete-time two-state Markovian chain in Chaps. 2 and 3 and multistate Markovian chain in Chaps. 4 and 5.

Certain application examples demonstrate that the nuance of reward function leads to completely different optimality of the myopic policy. Therefore, some efforts were made to discover the relation between the form of reward function and the optimality of myopic policy. A unified framework was constructed under the myopic policy in terms of the regular reward function, characterized by three basic axioms—symmetry, monotonicity, and decomposability. A wide variety of RMAB problems can be treated according to this framework, including satellite networks with optical crosslinks and RF downlinks, wireless ad hoc networks, computer networks, and hybrid networks with both wireless and wireline components only if they confirm to the decomposable rule. For the homogeneous channels in Chap. 2, we established the optimality of the myopic policy when the reward function can be expressed as a regular function and when the discount factor is bounded by a closed-form threshold determined by the reward function. Furthermore, the regular function was extended to a generic function such that it covers a much larger range of utility functions, particularly the logarithmic function and the power function widely used in engineering problems. By distinguishing arms as the sensed set and non-sensed set at each time slot, we quantified the trade-off of exploration vs. exploitation in the decision process, and then derived sufficient condition for the optimality of myopic policy.

K. Wang, L. Chen, *Restless Multi-Armed Bandit in Opportunistic Scheduling*,
https://doi.org/10.1007/978-3-030-69959-8_6

In order to further obtain the asymptotically optimal performance of the RMAB in all the parameter space, we analyzed its indexability and Whittle index for two-state Markovian case by the fixed-point approach in Chap. 3 and multistate Markovian case in Chap. 5. In Chap. 3, we first derived the threshold structure of the single-arm policy. Based on this structure, the closed-form Whittle index was obtained for the case of negatively correlated channels, while the Whittle index for the positively correlated channel was much complicated for its uncertainty, particularly for certain regions below stationary distribution of the Markovian chain. Then this region was divided into deterministic regions and indeterministic regions with interleaving structure. In the deterministic regions, the evolution of dynamic system is periodic and then there exists an eigen matrix to depict this kind of evolving structure through which the closed-form Whittle can be derived. In the indeterministic regions, there does not exist an eigen matrix to depict its aperiodic structure. In practical scenarios, given computing precision, the Whittle index in those regions can be computed in the simply linear interpolation since the distribution of the deterministic and indeterministic regions appears in the interleaving form.

In Chap. 4, we consider the downlink scheduling problem of wireless communication system consisting of N homogenous channels and one user in which each channel is assumed to evolve as a multistate Markov process. At each time instant, one channel is allocated to the user according to imperfect state observation information, and some reward is accrued depending on the state of the chosen channel. The objective is to design a scheduling policy that maximizes the expected accumulated discounted reward over an infinite horizon. Mathematically, the proposed problem can be formulated into a RMAB problem that is PSPACE hard with exponential memory and computation complexity. One of feasible approaches is to consider the myopic policy (or greedy policy) that only focuses on maximizing immediate reward with linear complexity. Specifically, in this chapter, we carry out a theoretic analysis on the performance of myopic policy with imperfect observation, introduce MLR order to characterize the evolving structure of belief information, and establish a set of closed-form conditions to guarantee the optimality of the myopic scheduling policy in downlink channel allocation.

In Chap. 5, we revisit the opportunistic scheduling problem where a server opportunistically serves multiple classes of users under time-varying multistate Markovian channels. The aim of the server is to find an optimal policy minimizing the average waiting cost of users. Mathematically, the problem can be cast to a restless bandit one, and a pivot to solve restless bandit by index policy is to establish indexability. Despite theoretical and practical importance of the index policy, the indexability is still open for the opportunistic scheduling in the heterogeneous multistate channel case. To fill this gap, we mathematically propose a set of sufficient conditions on channel state transition matrix under which the indexability is guaranteed, and consequently, the index policy is feasible. We further develop a simplified procedure to compute the index by reducing the complexity from more than quadratic to linear. Our work consists of a small step toward solving the opportunistic scheduling problem in its generic form involving multistate Markovian channels and multi-class users.

From the system perspective, our analysis presented in this book provides insight on the following design trade-off in opportunistic scheduling, artificial intelligence, and optimization operation.

- Exploitation Versus Exploration: Due to hardware limitations and cost constraint of consuming resources, a scheduler may not be able to observe all channels in system simultaneously. A good strategy is thus required for intelligent channel selection to track the rapidly varying system state. The purpose of a good strategy is twofold: to find good channels for immediate reward and to gain statistical information on the system state for better opportunity tracking in the future. Thus, the optimal strategy should strike a balance between these two often conflicting objectives.
- Aggression Versus Conservation: Based on the imperfect observation outcomes, a scheduler needs to decide whether to act or not. An aggressive strategy may lead to excessive collisions while a conservative one may result in performance degradation due to overlooked opportunities. Therefore, the optimal strategy should achieve a trade-off between aggression and conservation.

6.2 Open Issues and Directions for Future Research

6.2.1 RMAB with Multiple Schedulers

In this book, we mainly focus on the decision-making process and different trade-off with only one decision-maker or scheduler. A natural research direction is to take the results obtained in the book as a building block to further explore the opportunistic scheduling scenario with multiple schedulers. The key research challenge for multiple schedulers is how to coordinate them to utilize channels in a distributed fashion without or with little explicit network-level cooperation. A natural way to tackle this problem is to model the situation as a non-cooperative game among schedulers and to see whether the results obtained in this book can further be tailored in the new context.

6.2.2 RMAB with Switching Cost

Another aspect that may limit the performance of opportunistic scheduling mechanism is the channel switching cost. In current wireless devices or networks, the channel switching introduces a cost in terms of delay, packet loss, and protocol overhead. Hence, an efficient channel scheduling policy should avoid frequently channel switching, unless necessarily. In the context of RMAB, this problem can be mapped into the generic RMAB problem with switching cost between arms. More systematical works are called for to provide more in-depth insight on this problem.

It is important to note that the generic MAB with switching cost is NP-hard, and there does not exist any optimal index policy. More specifically, the introduction of switching cost renders not only the Gittins index policy suboptimal but also makes the optimal policy computationally prohibitive. Given such difficulties, we envision to tackle the problem from the following aspects:

- Looking for suboptimal policy with bounded efficiency loss compared to the optimal policy;
- Developing heuristic policy achieving a trade-off between optimality and complexity;
- Deriving optimal policy in a subset of scenarios or designing asymptotically optimal policy.

6.2.3 RMAB with Correlated Arms

Another practical extension is to consider the correlated channels, i.e., the Markov chains of different channels can be correlated. This problem can be cast into the RMAB problem with correlated arms. The introduction of the correlation among arms makes the trade-off between exploration and exploitation more sophisticated as sensing a channel can not only reveal the state of the sensed channel but also provide information on other channels as they are not entirely independent. How to characterize the trade-off in this new context and how to design efficient channel access policy are of course pressing research topics in this direction.

Index

K. Wang, L. Chen, *Restless Multi-Armed Bandit in Opportunistic Scheduling*,
https://doi.org/10.1007/978-3-030-69959-8

Printed in the United States
by Baker & Taylor Publisher Services